Linear Differential and Difference Equations:

A Systems Approach for Mathematicians and Engineers

The direction in which education starts a man will determine [his] future life.

Plato, *The Republic, Book IV*

Mathematics possesses not only truth, but supreme beauty—a beauty cold and austere, like that of sculpture, and capable of stern perfection, such as only great art can show.

Bertrand Russell in *the Principles of Mathematics*

Talking of education, "People have now a-days" (said he) "got a strange opinion that every thing should be taught by lectures. Now, I cannot see that lectures can do so much good as reading the books from which the lectures are taken. I know nothing that can be best taught by lectures, expect where experiments are to be shewn. You may teach chymestry by lectures. — You might teach making of shoes by lectures!"

James Boswell: *Life of Samuel Johnson, 1766*

ABOUT OUR AUTHOR

Roy Michael Johnson, Senior Lecturer in Mathematics at the University of Paisley, graduated with a B.Sc. (Honours) in mathematics from the University of Bristol in 1956. He is a Chartered Mathematician and a Fellow of the Institute of Mathematics and its Applications. After leaving university he worked for the De Havilland Aircraft Company at Hatfield, initially on a graduate training course, and later as an aerodynamicist on comparative performance studies of civil aircraft. He moved to Hawker Siddeley Dynamics in 1958 where his work as a dynamics engineer included design and development of missile guidance and control systems. In 1961 he was appointed Senior Dynamics Engineer with responsibilities for the development of new projects.

This long period in industry was to reinforce his teaching ability when he moved into education, joining Dundee College of Technology in 1964 as Lecturer in mathematics. In 1968 he became Lecturer in the same subject at Paisley, where his duties included development of continuous systems simulation, with special responsibility for all matters related to engineering mathematics. As industrial consultant to the National Engineering Laboratory, East Kilbride, for a number of years, he advised on problems related to vibrations in mechanical systems.

He now lectures to final year undergraduates of the B.Sc. Mathematical Sciences courses with specialisation in Control Theory and Three-dimensional Geometry. His recent research and publications are in the field of applications of geometry to graphics systems.

Mike Johnson is also the author of *Calculus* (Ellis Horwood Limited, 1987), rewritten and updated for Albion Publishing in 1995 and, with I.A. Huntley, of *Linear and Non-Linear Differential Equations* (Ellis Horwood Limited, 1983).

Linear Differential and Difference Equations:

A Systems Approach for Mathematicians and Engineers

R.M. Johnson, BSc(Hons), CMath, FIMA
Senior Lecturer
Department of Mathematics and Statistics
University of Paisley
Paisley

Albion Publishing
Chichester

First published in 1997 by
ALBION PUBLISHING LIMITED
International Publishers
Coll House, Westergate, Chichester, West Sussex, PO20 6QL England

British Library Cataloguing in Publication Data
A catalogue record of this book is available from the British Library

ISBN 1-898563-12-8

Printed in Great Britain by Hartnolls, Bodmin, Cornwall

CONTENTS

PART I CONTINUOUS SYSTEMS

PART II DISCRETE SYSTEMS

PREFACE

This textbook is a major revision of the book *Theory and Applications of Linear Differential and Difference Equations* first published in 1984 in the Ellis Horwood Series on Mathematics and its Applications. The motivation for the revision is based on the experience and perception of the original text by students at Paisley where the book was standard material for both mathematics and engineering courses.

The major changes in this new edition involve numerous simplifications, the omission of unused background material, the addition of a considerable number of graded examples and the inclusion of an additional section on digitally controlled feedback systems. The net result is a textbook which remains attractive to engineering students of all disciplines but is friendly to mathematics students. The necessary fundamental theory is not complicated by strange applications although a range of applications are developed once sufficient mathematics has been covered.

The textbook provides a compact treatment of linear differential equations and linear difference equations using transform techniques. It is easy to read and does not require a strong mathematical background. The level of presentation is approach to second and third year undergraduate students of mathematics and engineering. Additionally the book provides a useful starting point for those wishing to progress to advanced studies in transform mathematics, control theory and signal analysis. It offers to mathematicians an appreciation of how engineers use transforms and will simultaneously appeal to mathematically-minded engineers.

The transform techniques used in the book are developed to encourage readers to think in terms of transfer functions and block diagrams rather than equations, that is to adopt an algebraic approach to calculus problems. An important relationship between the transform variables and frequency is established and the book is probably unique in the way it uses frequency domain analysis to highlight the similarities between differential and difference equations. Examples are chosen from the fields of electrical, mechanical, civil and control engineering.

PART I (Chapters 10-4) covers continuous systems. Chapter 1 provides a brief introduction to Fourier series and Fourier transforms leading to Laplace transforms and emphasises the relationship between frequency and the Laplace variable. Readers who simply require the Laplace transform as a tool to solve linear differential equations may omit Chapter 1 and use the definition of the Laplace transform given in Chapter 2 as their starting point. Transfer functions and block diagrams are introduced in Chapter 2 along with combinations of systems including feedback systems. Chapter 3 deals with systems with oscillating inputs and introduces Bode diagrams and analog filters. Together Chapters 2 and 3 cover a course on the application of Laplace transforms to linear differential equations but the block diagram approach allows many short cuts whereby steady solutions may be determined without having to obtained a complete solution. In Chapter 4 delayed functions, periodic functions and systems with discontinuous inputs are considered. One example uses a form of sampler and provides a link with Part II.

PART II (Chapters 5-7) covers discrete systems, in particular digital systems. Chapter 5 uses an idealised sampling device to lead to the definition of the *z*-transform of a sequence. Properties of the *z* transform are developed and in Chapter 6 these are applied to the solution of linear difference equations. Chapter 6 also highlights the similarities in the methods of solution used for differential and difference equations when transfer functions and block diagrams are employed. Steady solutions to difference equations are obtained by shortcuts similar to those used for continuous systems. Digital filters are introduced in Chapter 7 and simple design algorithms are established so that the performance of a given analog filter may be copied. The final section deals with the compensation of a feedback system by means of a digital filter. Chapter 7 will be particularly useful for those whose background has been mainly in continuous systems.

The "dot" notation, $x \equiv dx/dt$, is used throughout the book and the symbol j represents $\sqrt{-1}$; other notations are defined as they occur in the text. All system inputs are taken to be zero for $t<0$.

Acknowledgements

The author is indebted to the many students at the University of Paisley who, in recent years, have been on the receiving end of much of the material contained in this textbook. Particular thanks are expressed to Anne Wylie for the considerable task of typing and preparing the manuscript in camera-ready format.

Finally, I am grateful to Ellis Horwood and his colleagues in Albion Publishing Limited for valuable assistance and encouragement throughout this project.

R.M. Johnson
Paisley, 1997

1

An Approach to the Laplace Transform

1.1 INTRODUCTION

The Laplace transform is a powerful mathematical tool for problems arising from the study of continuous systems. The term "continuous systems" is taken to imply systems which can be modelled by ordinary differential equations, for example

(i) a control system which positions a missile fin to achieve a certain lateral acceleration,

(ii) a crane where the position of the load is controlled by the application of hydraulic motors,

(iii) a structure subject to vibration.

The variables in these examples, position, acceleration, pressure, force, displacement are *continuous* variables which can take any value within some specified range.

In Part 1 of this book we will apply Laplace transforms to *linear* continuous systems, that is systems described by linear differential equations. This can be done by accepting the mathematical definition of a Laplace transform as a starting point and turning directly to Chapter 2. This preliminary chapter approaches the idea of a Laplace transform by considering the frequency characteristics of a function of time, and attempts to show the important relationship between the Laplace variable and frequency.

1.2 THE FOURIER SERIES OF A PERIODIC FUNCTION

A periodic function $f(t)$ satisfying certain conditions may be expressed as an infinite series which is a linear combination of sine and cosine functions whose frequencies are multiples of the fundamental frequency $w_0 = \frac{2\pi}{L}$, where L is the period of $f(t)$. The infinite series is known as the **Fourier Series Expansion** of $f(t)$ and takes the form

$$f(t) = \frac{1}{2}a_0 + \sum_{n=1}^{\infty} (a_n \cos nw_0t + b_n \sin nw_0t) \tag{1.1}$$

where the constants a_n and b_n are given by

$$a_n = \frac{w_0}{\pi} \int_{-\pi/w_0}^{\pi/w_0} f(t) \cos nw_0t \; dt \tag{1.2}$$

$$b_n = \frac{w_0}{\pi} \int_{-\pi/w_0}^{\pi/w_0} f(t) \sin nw_0t \; dt \tag{1.3}$$

$$\left(\text{note that } \frac{\pi}{w_0} = \frac{L}{2}\right).$$

Assuming that the Fourier series expansion exists for a given function $f(t)$, then equations (1.2) and (1.3) follow immediately from the orthogonality of the functions $\{\sin nw_0t, \cos nw_0t\}$ over the interval $\left[-\frac{\pi}{w_0}, \frac{\pi}{w_0}\right]$,

i.e. when $n \neq m$, m and n integers

$$\int_{-\pi/w_0}^{\pi/w_0} \cos nw_0t \cos mw_0t \, dt = 0$$

$$\int_{-\pi/w_0}^{\pi/w_0} \sin nw_0t \sin mw_0t \, dt = 0$$

and for all integers, m, n

$$\int_{-\pi/w_0}^{\pi/w_0} \sin nw_0t \cos nw_0t \, dt = 0.$$

Sufficient conditions for the existence of the series (1.1) are that the function $f(t)$ is bounded and has a finite number of discontinuities and a finite number of maxima and minima in the interval $\left[-\frac{\pi}{w_0}, \frac{\pi}{w_0}\right]$. If $t = t_1$ is a point where $f(t)$ is continuous, then the series converges to $f(t_1)$: if $t = t_2$ is a point where $f(t)$ is discontinuous, then the series converges to $\frac{1}{2}\{f(t_2^+) + f(t_2^-)\}$. For a comprehensive treatment of Fourier series see Kreysig (1993).

Example 1.1

Obtain the Fourier series expansion of the function

$$f(t) = \begin{cases} 1, & 0 \leq t < \pi \\ 0, & \pi \leq t < 2\pi \end{cases}, \quad f(t + 2\pi) = f(t).$$

The function is shown in Figure 1.1.

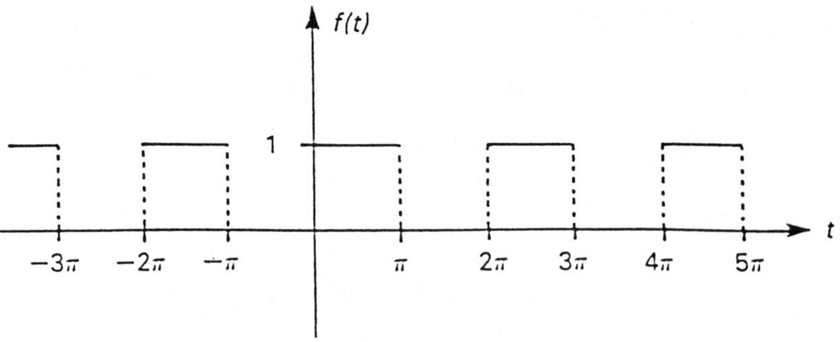

Fig. 1.1

The period of the function is $L = 2\pi$, therefore $w_0 = 1$.

Equations (1.2) and (1.3) give

$$a_0 = \frac{1}{\pi}\int_{-\pi}^{\pi} f(t)\,dt = \frac{1}{\pi}\int_{0}^{\pi} dt = 1$$

$$a_n = \frac{1}{\pi}\int_{-\pi}^{\pi} f(t)\cos nt\,dt \quad \text{for } n = 1, 2, 3, \ldots,$$

$$= \frac{1}{\pi}\int_{0}^{\pi} \cos nt\,dt = 0$$

$$b_n = \frac{1}{\pi}\int_{-\pi}^{\pi} f(t)\sin nt\,dt \quad \text{for } n = 1, 2, 3, \ldots,$$

$$= \frac{1}{\pi}\int_{0}^{\pi} \sin nt\,dt = \begin{cases} 2/\pi n\,, & \text{when } n \text{ is odd} \\ 0\,, & \text{when } n \text{ is even} \end{cases}.$$

Substituting these results into equation (1.1) gives the required series,

$$f(t) = 0.5 + \frac{2}{\pi}\left\{\sin t + \frac{\sin 3t}{3} + \frac{\sin 5t}{5} + \ldots\ldots\ldots\right\}.$$

Note that the above series converges to 1 for $0 < t < \pi$, converges to 0 for $-\pi < t < 0$ and converges (clearly) to 0.5 when $t = 0$. For example, putting $t = \frac{\pi}{2}$,

$$1 = 0.5 + \frac{2}{\pi}\left\{1 - \frac{1}{3} + \frac{1}{5} - \frac{1}{7} + \ldots\ldots\right\},$$

$$\text{i.e.} \quad \sum_{r=0}^{\infty} \frac{(-1)^r}{(2r+1)^2} = \frac{\pi}{4}.$$

Example 1.2

The function $f(t) = t$ is defined in the interval $0 \le t \le l$. Obtain an infinite series expansion of $f(t)$ of the form

$$f(t) = \frac{1}{2}a_0 + \sum_{n=1}^{\infty} a_n \cos\frac{n\pi}{l}t\,, \quad 0 \le t \le l.$$

Hence show that

$$\sum_{r=0}^{\infty} \frac{1}{(2r+1)^2} = \frac{\pi^2}{8}.$$

We consider the *even* function $f_E(t)$ defined as follows

$$f_E(t) = \begin{cases} t & , \quad 0 \le t < l \\ -t & , \quad -l \le t < 0 \end{cases} \quad , \; f_E(t + 2l) = f_E(t) \quad .$$

The function $f_E(t)$, shown in Figure 1.2(a), is a periodic function of period $2l$ and coincides with the given function $f(t)$ in the interval $0 \le t \le l$. Since it is an even function, $f_E(t)$ will have a Fourier series expansion which contains only cosine terms.

$$\text{i.e.} \qquad f_E(t) = \frac{1}{2} a_0 + \sum_{n=1}^{\infty} a_n \cos n w_0 t \ , \quad w_0 = \frac{\pi}{l} \ .$$

Therefore the function $f(t)$ will be represented by this series in the interval $0 \le t \le l$.

Equation (1.2) gives

$$a_0 = \frac{1}{l} \int_{-l}^{l} f_E(t)\ dt$$

$$= \frac{2}{l} \int_0^l t\ dt, \quad \text{since } f_E(t) \text{ is even,}$$

$$= l$$

$$a_n = \frac{1}{l} \int_{-l}^{l} f(t) \cos \frac{n\pi}{l} t\ dt\ , \qquad n = 1, 2, 3, \ldots\ldots$$

$$= \frac{2}{l} \int_0^l t \cos \frac{n\pi}{l} t\ dt$$

$$= \frac{2}{l} \left\{ \frac{l^2}{n\pi} \sin n\pi + \frac{l^2}{n^2\pi^2} (\cos n\pi - 1) \right\} .$$

Therefore
$$a_n = \begin{cases} 0 & , \quad n = 2, 4, 6, \ldots\ldots \\ \dfrac{-4l}{n^2\pi^2} & , \quad n = 1, 3, 5, \ldots\ldots \end{cases}$$

and
$$f(t) = \frac{l}{2} - \frac{4l}{\pi^2} \left\{ \cos \frac{\pi}{l} t + \frac{1}{3^2} \cos \frac{3\pi}{l} t + \frac{1}{5^2} \cos \frac{5\pi}{l} t + \ldots \right\} .$$

Setting $t = 0$ gives

$$0 = \frac{l}{2} - \frac{4l}{\pi^2} \left\{ 1 + \frac{1}{3^2} + \frac{1}{5^2} + \ldots\ldots \right\} ,$$

$$\text{i.e.} \qquad \sum_{r=0}^{\infty} \frac{1}{(2r+1)^2} = \frac{\pi^2}{8} .$$

Note that if we require a *sine* series for $f(t) = t\ ,\ 0 \le t \le l$, then it is necessary to consider the periodic *odd* function.

$$f_0(t) = t\ , \quad -l \le t \le l\ , \; f_0(t + 2l) = f_0(t) .$$

The Fourier series expansion of $f_0(t)$ contains only sine terms, i.e. $a_n = 0$ for $n = 0, 1, 2, 3, \ldots$.

See Figure 1.2(b) and Problem 3.

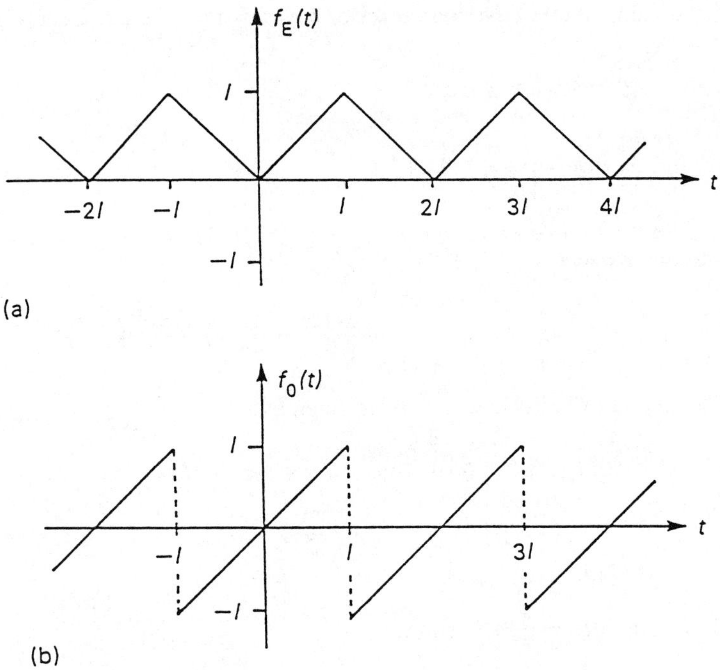

Fig. 1.2 (a) Even extension of $f(t)$, (b) Odd extension of $f(t)$.

Problems

1. The periodic function $f(t)$ is defined as follows:

$$f(t) = \begin{cases} 0 \ , & -\pi \le t < -\frac{\pi}{2} \\ 2 \ , & -\frac{\pi}{2} \le t \le \frac{\pi}{2} \\ 0 \ , & \frac{\pi}{2} < t \le \pi \end{cases} \ , \quad f(t+2\pi) = f(t) \ .$$

Show that the Fourier series expansion for $f(t)$ is

$$f(t) = 1 + \frac{4}{\pi}\left(\cos t - \frac{1}{3}\cos 3t + \frac{1}{5}\cos 5t - \frac{1}{7}\cos 7t + \dots\right).$$

2. Obtain the Fourier series expansion of the function

$$f(t) = \begin{cases} \pi + t \ , & -\pi \le t < 0 \\ \pi - t \ , & 0 \le t < \pi \end{cases} \ , \quad f(t+2\pi) = f(t) \ .$$

3. The function $f(t) = t$ is defined on the interval $0 \le t \le l$. Obtain an infinite series expansion of $f(t)$ in the form

$$f(t) = \sum_{n=1}^{\infty} b_n \sin \frac{n\pi}{l} t \ .$$

4. By consideration of the Fourier series for $f(t) = t^2, -\pi \le t \le \pi$, show that

(i) $\dfrac{\pi^2}{6} = 1 + \dfrac{1}{4} + \dfrac{1}{9} + \dfrac{1}{16} + \ldots$

(ii) $\dfrac{\pi^2}{12} = 1 - \dfrac{1}{4} + \dfrac{1}{9} - \dfrac{1}{16} + \ldots$

5. Obtain the Fourier series for the even function $f(t) = e^{-a|t|}$, $a > 0, -L \le t \le L$ and deduce that

$$\sum_{r=1}^{\infty} \left\{a^2 + (2r-1)^2 \frac{\pi^2}{L^2}\right\}^{-1} = \frac{L}{4a}\left\{\frac{1 - e^{-aL}}{1 + e^{-aL}}\right\}.$$

6. The output of half-wave rectifier is given by

$$f(t) = \begin{cases} E \sin \dfrac{2\pi}{T} t & , \quad 0 \le t < \dfrac{T}{2} \\ 0 & , \quad \dfrac{T}{2} \le t < T \end{cases} , \quad f(t + T) = f(t) .$$

Show that $f(t) = \dfrac{E}{\pi} + \dfrac{E}{2} \sin \dfrac{2\pi}{T} t - \dfrac{2E}{\pi}\left\{\dfrac{\cos \frac{4\pi}{T} t}{1.3} + \dfrac{\cos \frac{8\pi}{T} t}{3.5} + \ldots\ldots\right\}.$

Answers To Problems

2. $\dfrac{\pi}{2} + \dfrac{4}{\pi}\left\{\cos t + \dfrac{1}{9}\cos 3t + \dfrac{1}{25}\cos 5t + \ldots\ldots\right\}.$

3. $\dfrac{2l}{\pi}\left\{\sin \dfrac{\pi}{l} t - \dfrac{1}{2}\sin \dfrac{2\pi}{l} t + \dfrac{1}{3}\sin \dfrac{3\pi}{l} t - \ldots\ldots\right\}, \quad 0 \le t \le l.$

5. $\dfrac{1}{aL}(1 - e^{-aL}) + \displaystyle\sum_{n=1}^{\infty} (1 - (-1)^n e^{-aL})\left(a^2 + \dfrac{n^2\pi^2}{L^2}\right)^{-1} \cos \dfrac{n\pi}{L} t .$

1.3 COMPLEX FORM OF FOURIER SERIES

The sequence of numbers $\{a_0, a_1, a_2, \ldots, b_1, b_2, \ldots\}$ is an alternative way of defining the function $f(t)$ since given these numbers we can reconstruct the function using the Fourier series, equation (1.1).

Writing the series as $f(t) = \frac{1}{2} a_0 + \sum_{n=1}^{\infty} r_n \cos (nw_0 t + \phi_n)$ where $r_n = \sqrt{a_n^2 + b_n^2}$, $n = 1, 2, 3, \ldots\ldots$, each number r_n provides a measure of the contribution to the periodic function $f(t)$ from the frequency nw_0. This information can be presented graphically in the form of a **discrete amplitude spectrum**, see Figure 1.3. In particular, the discrete amplitude spectrum for Example 1.1 is shown in Figure 1.4.

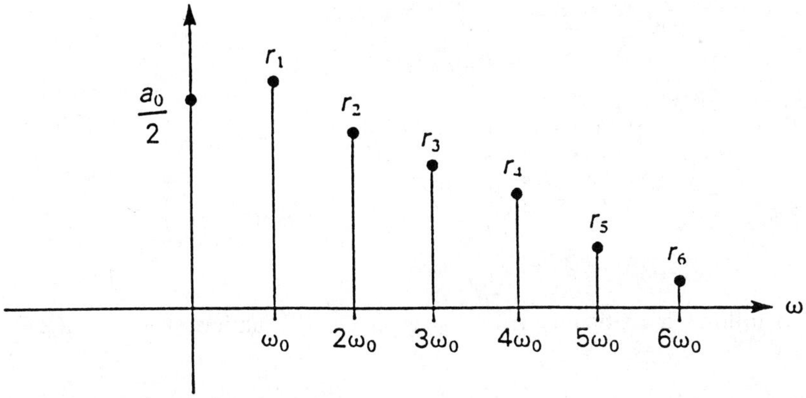

Fig. 1.3

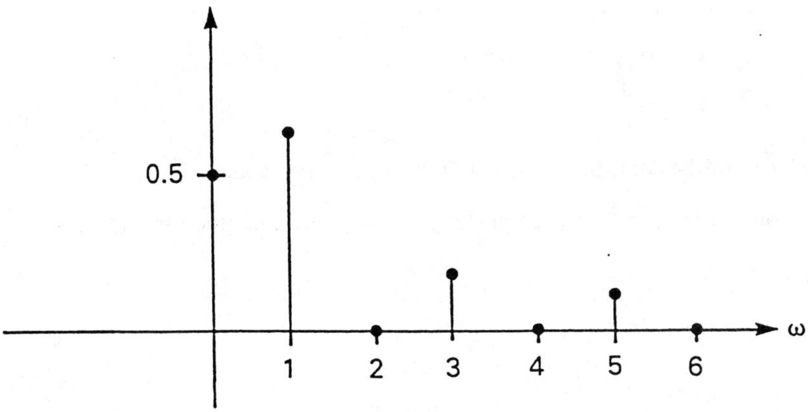

Fig. 1.4

An alternative, and more usual, representation of the discrete amplitude spectrum is related to the *complex form* of the Fourier series expression.

Making the substitutions

$$\cos nw_0t = \frac{1}{2}(e^{jnw_0t} + e^{-jnw_0t})$$

$$\sin nw_0t = \frac{1}{2j}(e^{jnw_0t} - e^{-jnw_0t})$$

where $j^2 = -1$, equation (1.1) becomes

$$f(t) = \frac{1}{2}a_0 + \frac{1}{2}\sum_{n=1}^{\infty}\{(a_n - jb_n)e^{jnw_0t} + (a_n + jb_n)e^{-jnw_0t}\} \qquad (1.4)$$

If we define
$$c_0 = \frac{1}{2} a_0$$
$$c_n = \frac{1}{2}(a_n - jb_n),\ n = 1, 2, 3, \ldots,$$
$$c_{-n} = \frac{1}{2}(a_n + jb_n),\ n = 1, 2, 3, \ldots,$$

then the equation (1.4) can be written in the *complex form*

$$f(t) = \sum_{n=-\infty}^{n=\infty} c_n e^{jnw_0 t} \tag{1.5}$$

where

$$c_n = \frac{w_0}{2\pi} \int_{-\pi/w_0}^{\pi/w_0} f(t)\, e^{-jnw_0 t}\, dt\ . \tag{1.6}$$

Equation (1.6) follows immediately from equation (1.2), equation (1.3) and the definition of c_n.

The sequence of numbers $\{c_n, n = 0, \pm 1, \pm 2, \pm 3, \ldots\}$ is an alternative method of defining the function $f(t)$, since the function can be reconstructed from these coefficients using the complex form Fourier series, equation (1.5).

Note that, if $f(t)$ is a real function, then

$$c_{-n} = \bar{c}_n \text{ and}$$

$$|c_n| = |c_{-n}| = \frac{1}{2}\sqrt{a_n^2 + b_n^2} = \frac{1}{2} r_n$$

i.e. $|c_n|$ is one half of the contribution to $f(t)$ from the frequency nw_0.

This information can be represented graphically as a discrete amplitude spectrum, see Figure 1.5.

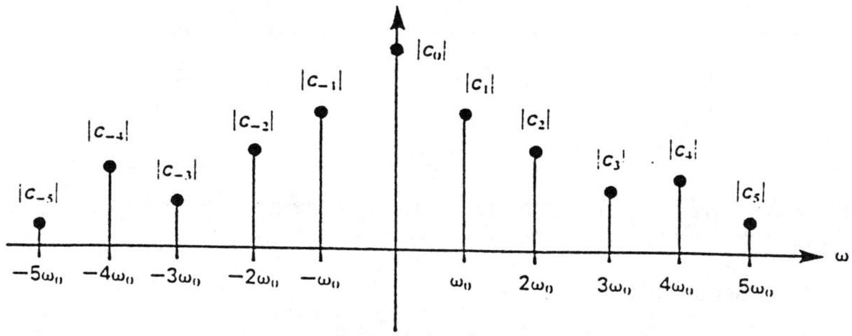

Fig. 1.5 Discrete amplitude spectrum for real periodic function.

Note that the discrete amplitude spectrum is symmetrical about $w = 0$ for a real function $f(t)$. The amplitude contribution to $f(t)$ from the frequency nw_0 is obtained by adding the height of the spectrum at nw_0 and at the "negative frequency" $-nw_0$

i.e.
$$r_n = |c_n| + |c_{-n}|$$
$$= 2|c_n| \text{ for a real function.}$$

Example 1.3

Obtain the complex form Fourier series for the function

$$f(t) = \begin{cases} 1 & , \quad 0 \le t < \pi \\ 0 & , \quad \pi \le t < 2\pi \end{cases} \quad , \; f(t+2\pi) = f(t)$$

and sketch its discrete amplitude spectrum. (This function was considered in example 1.1.)

From equation (1.6)

$$c_n = \frac{1}{2\pi} \int_{-\pi}^{\pi} f(t)\, e^{-jnt}\, dt$$

$$= \frac{1}{2\pi} \int_{0}^{\pi} e^{-jnt}\, dt$$

$$= -\frac{1}{2\pi jn} \left[e^{-jnt} \right]_0^{\pi} \quad , \quad n \ne 0$$

$$= \frac{j}{2\pi n} \left(e^{-jn\pi} - 1 \right), \quad n \ne 0$$

$$c_0 = \frac{1}{2\pi} \int_0^{\pi} dt = \frac{1}{2} .$$

Therefore

$$c_n = \begin{cases} \dfrac{1}{2} & , \quad n = 0 \\ \dfrac{-j}{\pi n} & , \quad n = \pm 1 \,,\, \pm 3 \,,\, \pm 5 \,,\, \ldots\ldots \\ 0 & , \quad n = \pm 2 \,,\, \pm 4 \,,\, \pm 6 \,,\, \ldots\ldots \end{cases}$$

The complex form Fourier series can be written as

$$f(t) = \frac{1}{2} - \frac{j}{\pi} \sum_{n=-\infty}^{n=\infty} \frac{e^{j(2n-1)t}}{(2n-1)} .$$

The discrete amplitude spectrum is shown in Figure 1.6.

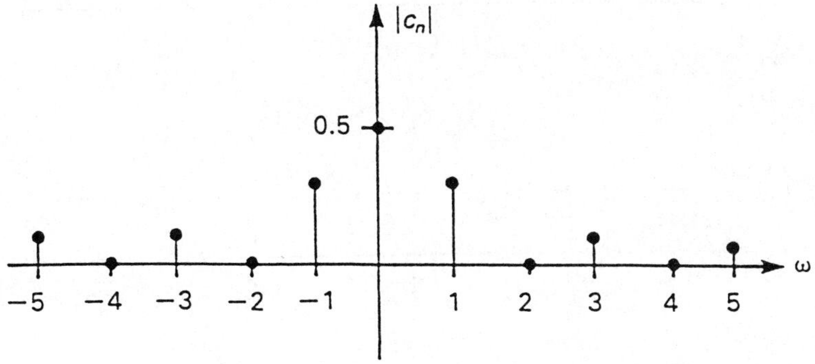

Fig. 1.6 Discrete amplitude spectrum for Example 1.3

Problems

1. Show that the complex form of the Fourier series of the function $f(t) = e^t$ when $-\pi < t < \pi, f(t + 2\pi) = f(t)$ is given by

$$f(t) = \frac{2 \sinh \pi}{\pi} \sum_{n=-\infty}^{n=\infty} (-1)^n \frac{1 + jn}{1 + n^2} e^{jnt} .$$

2. From the series in Problem 1, deduce that

$$f(t) = \frac{2 \sinh \pi}{\pi} \left\{ \frac{1}{2} - \frac{1}{2}(\cos t - \sin t) + \frac{1}{5}(\cos 2t - 2 \sin 2t) - \ldots\ldots \right\} .$$

3. Given that $f(t) = \sum_{n=-\infty}^{n=\infty} c_n e^{jnwt}$, show that when $f(t)$ is an odd function c_n is purely imaginary, and when $f(t)$ is an even function c_n is real.

1.4 EXTENSION TO NON-PERIODIC FUNCTIONS : THE FOURIER TRANSFORM

We consider a non-periodic function as the limit of a periodic function when the period is allowed to increase without bound. For example, consider the function

$$f(t) = \begin{cases} 1 & \text{when} \quad 0 \le t < 1 \\ 0 & \text{otherwise} \end{cases} ,$$

which is shown in Figure 1.7(a). This function can be considered as $\lim_{T \to \infty} f_p(t)$ where $f_p(t)$ is the periodic function shown in Figure 1.7(b) and defined by

$$f_p(t) = \begin{cases} 1 & \text{when} \quad 0 \le t < 1 \\ 0 & \text{when } 1 \le t < T \end{cases} , \quad f_p(t + T) = f_p(t).$$

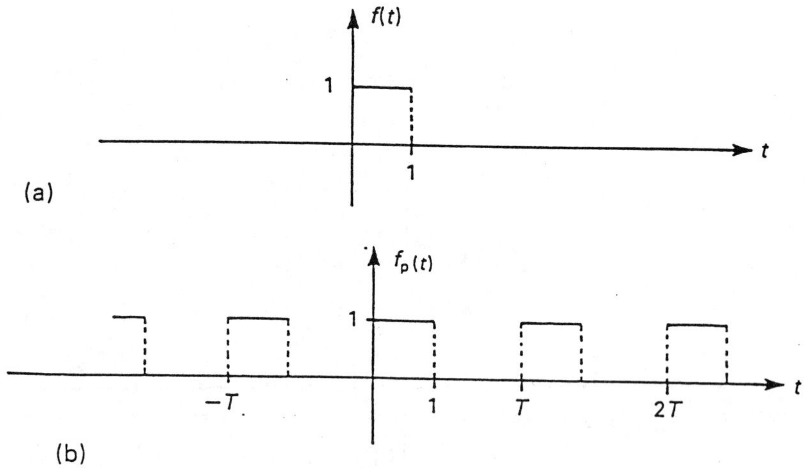

Fig. 1.7 $f(t) = \lim_{T \to \infty} \{f_p(t)\}$

In general we can express a periodic function $f_p(t)$, with period T, as complex form Fourier series according to equations (1.5) and (1.6). Substituting for c_n in equation (1.5) gives

$$f_p(t) = \sum_{n=-\infty}^{\infty} \left\{ \frac{w_0}{2\pi} \int_{-\pi/w_0}^{\pi/w_0} f(u)e^{-jnw_0 u}\, du \right\} e^{jnw_0 t} \qquad (1.7)$$

where $w_0 = \frac{2\pi}{T}$ and the dummy variable in equation (1.6) has been changed from t to u to avoid confusion.

Defining $w = nw_0$ to be a *continuous* frequency variable and $\Delta w = w_0$, the difference between adjacent ordinates on the discrete amplitude spectrum of $f_p(t)$, equation (1.7) can be written

$$f_p(t) = \frac{1}{2\pi} \sum_{n=-\infty}^{\infty} \left\{ \int_{-\pi/\Delta w_0}^{\pi/\Delta w_0} f(u)e^{-jwu}\, du \right\} e^{jwt}\, \Delta w \ .$$

Now letting the period $T \to \infty$, i.e. $w_0 = \Delta w \to 0$ and using a fundamental theorem of integral calculus (a definite integral is the limit of a summation) we obtain

$$f(t) = \frac{1}{2\pi} \int_{-\infty}^{\infty} \left\{ \int_{-\infty}^{\infty} f(u)e^{-jwu}\, du \right\} e^{jwt}\, dw \qquad (1.8)$$

This can be written

$$f(t) = \frac{1}{2\pi} \int_{-\infty}^{\infty} F(jw)e^{jwt}\, dw \qquad (1.9)$$

where

$$F(jw) = \int_{-\infty}^{\infty} f(t)e^{-jwt}\, dt \ , \qquad (1.10)$$

(changing the dummy variable back to t).

Equation (1.10) defines the function $F(jw)$ which is called the **Fourier transform** of $f(t)$. Equation (1.9) is the **inversion formula**, i.e. $f(t)$ is the **inverse Fourier transform** of $F(jw)$. The equations (1.9) and (1.10) are called a **Fourier transform pair** and together are equivalent to equation (1.8) which is known as the **Fourier integral formula**. Other definitions for Fourier transform and inversion formula are given in the literature but taken *together* will reduce to equation (1.8). (An alternative is to change the frequency variable to $v = \frac{w}{2\pi}$ (Hertz) which makes equations (1.9) and (1.10) the same apart from the sign of j). For a more detailed treatment of the Fourier transform see Oppenheim and Willsky (1983).

The Fourier integral formula is the equivalent for non-periodic functions of the Fourier series expansion, and the function $F(jw)$ corresponds to the set of Fourier coefficients $\{c_n;\ n = 0, \pm 1, \pm 2, \ldots\}$. $F(jw)$ is a function of w and is referred to as the **spectrum** of $f(t)$. The **amplitude spectrum** $|F(jw)|$ is now a continuous function of w, see Figure 1.8. By writing $e^{-jwt} = \cos wt - j \sin wt$ in equation (1.10), it is easy to show that $|F(jw)| = |F(-jw)|$, i.e. the amplitude spectrum is an *even* function of w.

The function $F(jw)$ is an alternative way of representing the function $f(t)$ since, given $F(jw)$, we can reconstruct the function $f(t)$ using the inversion formula. The spectrum $F(jw)$ is called the representation of $f(t)$ in the **frequency domain**. Assuming that the integral in equation (1.10) converges then $F(jw)$ is unique for a given $f(t)$. However, not all functions of $f(t)$ can be expressed in the form of the Fourier integral formula, i.e. the Fourier transform $F(jw)$ does not exist for certain functions of $f(t)$. *Sufficient* conditions for the existence of $F(jw)$ are:

(i) $f(t)$ has a finite number of maxima and minima and discontinuities in any finite interval,

(ii) $\int_{-\infty}^{\infty} |f(t)|\, dt$ is finite.

We define the **energy** of a function $f(t)$ as $\int_{-\infty}^{\infty} \{f(t)\}^2\, dt$ and the conditions (i) and (ii) include all useful functions with *finite* energy.

From equation (1.9) it is easy to show that the energy of $f(t)$ can also be expressed as

$$\frac{1}{\pi}\int_0^{\infty} |F(jw)|^2\, dw.$$

It follows that the energy of $f(t)$ contained in the frequency band $w_1 < w < w_2$ is

$$\frac{1}{\pi}\int_{w_1}^{w_2} |F(jw)|^2\, dw\,.$$

Many simple functions, however, do not possess finite energy and condition (ii) is not satisfied, e.g. $f(t) = 1, f(t) = \sin t$. We define the **power** of a function as energy per unit time, i.e.

$$\lim_{T\to\infty}\frac{1}{T}\int_{-T/2}^{T/2} \{f(t)\}^2\, dt\,.$$

Functions with *infinite* energy and *finite* power do not possess Fourier Transforms in terms of elementary functions, but in certain cases the transforms can be expressed in terms of impulse functions, see Example 1.5. For further information on power signals and energy signals see Gabel and Roberts (1973).

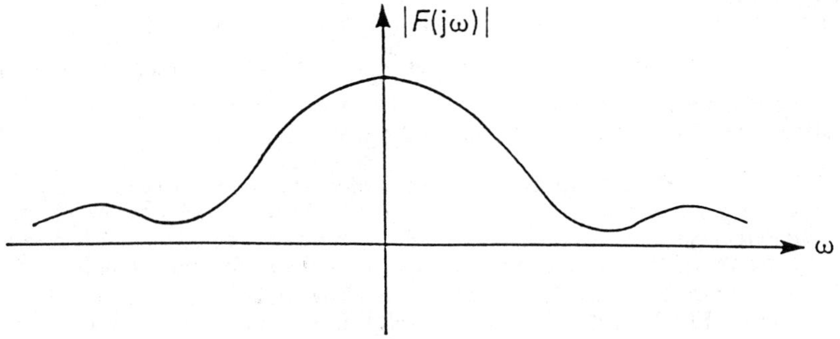

Fig. 1.8 Amplitude spectrum for real non-periodic function.

Example 1.4

Find the Fourier transform of the function shown in Figure 1.7(a)

i.e. $$f(t) = \begin{cases} 1 & \text{when} \quad 0 \le t < 1 \\ 0 & \text{otherwise} \end{cases} .$$

Hence show that

$$\int_0^{\infty} \frac{\sin w}{w}\, dw = \frac{\pi}{2}$$

$$\begin{aligned} F(jw) &= \int_{-\infty}^{\infty} f(t)\, e^{-jwt}\, dt = \int_0^1 e^{-jwt}\, dt \\ &= -\frac{1}{jw}\left[e^{-jwt}\right]_0^1 \\ &= \frac{1 - e^{-jw}}{jw} . \end{aligned}$$

Using the inversion formula, equation (1.9),

$$f(t) = \frac{1}{2\pi} \int_{-\infty}^{\infty} \left(\frac{1 - e^{-jw}}{jw}\right) e^{jwt}\, dw.$$

This integral representation of $f(t)$ will converge to the value of $f(t)$ at points where $f(t)$ is continuous, but at a point of discontinuity of $f(t)$ the integral will converge to $\frac{1}{2}\{f(t^+) + f(t^-)\}$.

Therefore, when $t = 0$, we have

$$\begin{aligned} \frac{1}{2}\{1 + 0\} &= \frac{1}{2\pi} \int_{-\infty}^{\infty} \left(\frac{1 - e^{-jw}}{jw}\right) dw \\ &= \frac{1}{2\pi} \int_{-\infty}^{\infty} \left(\frac{1 - \cos w}{jw}\right) dw + \frac{1}{2\pi} \int_{-\infty}^{\infty} \frac{\sin w}{w}\, dw \\ &= \frac{1}{\pi} \int_0^{\infty} \frac{\sin w}{w}\, dw , \end{aligned}$$

since the first integrand is an odd function of w and the second integrand is an even function of w.

Therefore,

$$\int_0^{\pi} \frac{\sin w}{w}\, dw = \frac{\pi}{2} .$$

(Note how the Fourier integral formula has been used to evaluate an improper integral in a similar way to the use of Fourier series to sum the infinite series in Examples 1.1 and 1.2.)

Example 1.5

We define the following functions.

(i) The *symmetric pulse function* $P_\alpha(t)$,

$$P_\alpha(t) = \begin{cases} 1 & \text{when } -\frac{\alpha}{2} \le t < \frac{\alpha}{2} \\ 0 & \text{otherwise} \end{cases} .$$

(ii) The *delayed pulse function* $P_\alpha(t-T)$,

$$P_\alpha(t-T) = \begin{cases} 1 & \text{when } T-\frac{\alpha}{2} \le t < T+\frac{\alpha}{2} \\ 0 & \text{otherwise} \end{cases} .$$

(iii) The *unit impulse function* $\delta(t)$,

$$\delta(t) = \lim_{\alpha \to 0} \frac{1}{\alpha} P_\alpha(t) .$$

(iv) The *delayed impulse function* $\delta(t-T)$,

$$\delta(t-T) = \lim_{\alpha \to 0} \frac{1}{\alpha} P_\alpha(t-T) .$$

Obtain the Fourier transforms of $P_\alpha(t - T)$ and $\delta(t)$. Deduce that the Fourier transform of the function $f(t) = 1$ is $2\pi\,\delta(w)$.

By considering the Fourier transform of $\delta(t - T) + \delta(t + T)$ obtain the Fourier transform of the function $\cos w_0 t$.

For the function $P_\alpha(t-T)$, the Fourier transform is given by

$$\int_{T-\frac{\alpha}{2}}^{T+\frac{\alpha}{2}} e^{-jwt}\,dt$$

$$= \frac{e^{-jw\left(T-\frac{\alpha}{2}\right)} - e^{-jw\left(T+\frac{\alpha}{2}\right)}}{jw}$$

$$= \frac{2}{w} e^{-jwT} \sin\frac{w\alpha}{2} .$$

Setting $T = 0$, the Fourier transform of $P_\alpha(t)$ is simply $\frac{2}{w}\sin\frac{w\alpha}{2}$ and the Fourier transform of $\delta(t)$ is given by

$$\lim_{\alpha \to 0} \left\{\frac{2}{\alpha w} \sin\frac{w\alpha}{2}\right\}$$

$$= \lim_{\theta \to 0} \left\{ \frac{\sin \theta}{\theta} \right\}, \quad \text{where } \theta = \frac{w\alpha}{2}$$
$$= 1.$$

Since the Fourier transform of $\delta(t)$ is 1 the inversion formula gives

$$\delta(t) = \frac{1}{2\pi} \int_{-\infty}^{\infty} e^{jwt}\, dw.$$

By simply interchanging t and the dummy variable w, we obtain after a sign change substitution

$$2\pi\, \delta(w) = \int_{-\infty}^{\infty} e^{-jwt}\, dt.$$

Therefore, the Fourier transform of the function $f(t) = 1$ is given by $2\pi\delta(w)$.

The Fourier transform of $\delta(t - \mathrm{T}) + \delta(t + \mathrm{T})$ is given by

$$\lim_{\alpha \to 0} \left\{ \frac{2}{\alpha w} e^{-jw\mathrm{T}} \sin \frac{w\alpha}{2} + \frac{2}{\alpha w} e^{jw\mathrm{T}} \sin \frac{w\alpha}{2} \right\}$$
$$= e^{jw\mathrm{T}} + e^{-jw\mathrm{T}}$$
$$= 2 \cos w\mathrm{T}.$$

The inversion formula gives

$$\delta(t - \mathrm{T}) + \delta(t + \mathrm{T}) = \frac{1}{2\pi} \int_{-\infty}^{\infty} 2 \cos w\mathrm{T} \,.\, e^{jwt}\, dw\,.$$

Again interchanging t and the dummy variable w, and writing $\mathrm{T} = w_0$ we obtain

$$\pi\, \{\delta(w - w_0) + \delta(w + w_0)\} = \int_{-\infty}^{\infty} \cos w_0 t \,.\, e^{-jwt}\, dt\,.$$

It follows that the Fourier transform of the function $\cos w_0 t$ is

$$\pi(\delta(w - w_0) + \delta(w + w_0)).$$

It is interesting to note how the *dual* nature of the Fourier transform pair allows the transforms of the functions 1 and $\cos w_0 t$ to be readily obtained. These functions do not satisfy the condition

$$\int_{-\infty}^{\infty} |f(t)|\, dt < \infty$$

and consequently their Fourier transforms contain non-elementaray functions. Impulse functions are entirely theoretical, i.e. they cannot be sketched or generated by physical equipment. The spectrum of $\cos w_0 t$ consists only of impulses "of strength π" at $w = \pm\, w_0$. This confirms the obvious fact that all the *power* of the function $\cos w_0 t$ is concentrated at the frequency w_0.

Problems

1. Use a result from Example 1.5 to deduce that the Fourier transform of the function $\frac{\sin t}{t}$ is $\pi \mathrm{P}_2(w)$.

2. Find the Fourier transform of the function $f(t) = e^{-a|t|}$, $a > 0$, and deduce that

$$\int_0^\infty \frac{\cos w}{a^2 + w^2}\, dw = \frac{\pi}{2a} e^{-a} .$$

3. Obtain the Fourier transform of the function

$$f(t) = \begin{cases} -e^{at} & \text{when } t < 0 \\ e^{-at} & \text{when } t > 0 \end{cases}, \quad a > 0.$$

Hence show that this Fourier transform $\rightarrow \frac{2}{jw}$ as $a \rightarrow 0$ and deduce that the Fourier transform of the *unit step function* $u(t)$,

$$u(t) = \begin{cases} 1 & t \geq 0 \\ 0 & t < 0 \end{cases},$$

is given by $\pi\delta(w) + \frac{1}{jw}$.

Answers to Problems

2. $\frac{2a}{a^2 + w^2}$.

3. $\frac{1}{a + jw} - \frac{1}{a - jw}$.

1.5 THE LAPLACE TRANSFORM

In Section 1.4, we saw the difficulties which arose when we attempted to obtain the Fourier transform of certain functions $f(t)$ which did not tend to zero as $t \rightarrow \infty$. For these functions the condition

$$\int_{-\infty}^{\infty} |f(t)|\, dt < \infty$$

is not satisfied and the integral $\int_{-\infty}^{\infty} f(t)\, e^{-jwt}\, dt$ does not converge. The Laplace transform can be considered as a modification of the Fourier transform to overcome the difficulties described above. If we multiply the function $f(t)$ by $e^{-\sigma|t|}$ with $\sigma > 0$ then for a large class of functions the integral

$$\int_{-\infty}^{\infty} f(t)\, e^{-\sigma|t|}\, e^{-jwt}\, dt$$

will converge for a suitable value of σ. Most physical problems can be defined so that $f(t)$ is zero for $t < 0$ and the modified integral can be written

$$\int_0^\infty f(t)\, e^{-(\sigma + jw)t}\, dt .$$

Using the notation of equation (1.10), we have

$$F(\sigma + jw) = \int_0^\infty f(t)\, e^{-(\sigma+jw)t}\, dt \qquad (1.11)$$

Defining the complex variable $s = \sigma + jw$, we obtain

$$F(s) = \int_0^\infty f(t)\, e^{-st} dt \qquad (1.12)$$

F(s) is called the **Laplace transform** of $f(t)$, and the complex variable s, whose imaginary part is the frequency variable, is known as the **Laplace variable**. Note that the Laplace transform of $f(t)$ is also the Fourier transform of $f(t)e^{-\sigma t}$, assuming that $f(t)$ is zero for $t < 0$.

Alternative notations for the transform pair are

$$F(s) = L\,\{f(t)\}\ ,\quad f(t) = L^{-1}\,\{F(s)\}$$

where L is the Laplace transform *operator* and L^{-1} is the *inverse operator*. An inversion formula for the Laplace transform exists but involves an integration in the complex s-plane, and it is more convenient to use tabulated transform pairs for the common functions, see Table 1.

Table 1 Laplace transforms of *f(t), f(t)* = 0 for *t* < 0

$$L\{f(t)\} = \int_0^\infty f(t)\, e^{-st}\, dt = F(s)$$

$f(t)$	$F(s)$
1	$\frac{1}{s}$
t	$\frac{1}{s^2}$
t^2	$\frac{2}{s^3}$
$\frac{t^{n-1}}{(n-1)!}$, $n = 1, 2, 3, \ldots$	$\frac{1}{s^n}$
$t\ f(t)$	$-\ F'(s)$
$u(t - T)$	$\frac{e^{-sT}}{s}$
$\delta(t)$	1

$\delta(t - T)$	e^{-sT}
$f(t - T)\ u(t - T)$	$e^{-sT}\ F(s)$
e^{-at}	$\dfrac{1}{(s + a)}$
$t\ e^{-at}$	$\dfrac{1}{(s + a)^2}$
$\cos\ wt$	$\dfrac{s}{s^2 + w^2}$
$\sin\ wt$	$\dfrac{w}{s^2 + w^2}$
$e^{-at}\cos\ wt$	$\dfrac{(s + a)}{(s + a)^2 + w^2}$
$e^{-at}\sin\ wt$	$\dfrac{w}{(s + a)^2 + w^2}$
$e^{-at}\ f(t)$	$F(s + a)$
$f'(t)$	$s\,F(s) - f(0)$
$f''(t)$	$s^2\ F(s) - s\,f(0) - f'(0)$
$f^{(n)}(t)$	$s^n\ F(s) - \sum_{r=1}^{n} s^{n-r} f^{(r-1)}(0)$
$\int_0^t f(u)\ du$	$\dfrac{F(s)}{s}$

NOTE:

(i) In each case it is assumed that s is such that the integral converges.

(ii) $f^{(n)}(t) \equiv \dfrac{d^n f}{dt^n}$.

Example 1.6

Obtain the Laplace transform of (i) the unit step function $u(t)$, (ii) the function e^{-at}, $a > 0$.

(i)
$$u(t) = \begin{cases} 1 & t \geq 0 \\ 0 & t < 0 \end{cases} .$$

$$\begin{aligned} \mathrm{L}\{u(t)\} &= \int_0^\infty e^{-st}\,dt \\ &= -\frac{1}{s}\,[e^{-st}]_0^\infty \\ &= \frac{1}{s}, \text{ assuming } \sigma > 0. \end{aligned}$$

(ii)
$$\begin{aligned} \mathrm{L}\{e^{-at}\} &= \int_0^\infty e^{-(a+s)t}\,dt \\ &= -\frac{1}{s+a}\,[e^{-(a+s)t}]_0^\infty \\ &= \frac{1}{s+a}, \text{ assuming } \sigma > -a. \end{aligned}$$

Example 1.7

(i) Writing $e^{jwt} = \cos wt + j \sin wt$ find the Laplace transforms of $\cos wt$ and $\sin wt$.

(ii) Obtain the inverse Laplace transform of $\dfrac{2-3s}{(s+2)(s^2+4)}$.

(i) Following the working of part (ii) in Example 1.6,

$$\mathrm{L}\{e^{jwt}\} = \frac{1}{s-jw} \qquad \text{assuming } \sigma > 0.$$

Taking real and imaginary parts,

$$\mathrm{L}\{\cos wt\} = \frac{s}{s^2+w^2}$$

$$\mathrm{L}\{\sin wt\} = \frac{w}{s^2+w^2} .$$

(ii) Using the method of partial fractions

$$\begin{aligned} \frac{2-3s}{(s+2)(s^2+4)} &= \frac{1}{(s+2)} - \frac{s+1}{(s^2+4)} \\ &= \frac{1}{(s+2)} - \frac{s}{(s^2+2^2)} - \frac{1}{2}\left(\frac{2}{s^2+2^2}\right) . \end{aligned}$$

Therefore $\mathrm{L}^{-1}\left\{\dfrac{2-3s}{(s+2)(s^2+4)}\right\} = e^{-2t} - \cos 2t - \dfrac{1}{2}\sin 2t$,

using the results in part (i) and the previous example.

Example 1.8

Show that $$\mathrm{L}\left\{\frac{df}{dt}\right\} = s\,\mathrm{L}\{f(t)\} - f(0).$$

$$\mathrm{L}\left\{\frac{df}{dt}\right\} = \int_0^\infty \frac{df}{dt}\, e^{-st}\, dt$$

$$= [e^{-st} f(t)]_0^\infty + s \int_0^\infty f(t)\, e^{-st}\, dt\ ,$$

(integration by parts).

Therefore $$\mathrm{L}\left\{\frac{df}{dt}\right\} = -f(0) + s\mathrm{L}\{f(t)\}\ ,$$ assuming that σ is large enough for $\lim_{t\to\infty} (e^{-st} f(t)) = 0.$

Problems

1. Obtain the Laplace transform of the function

$$u(t-1) = \begin{cases} 1 & \text{when } t \geq 1 \\ 0 & \text{when } t < 1 \end{cases}\ ,$$

and hence write down the Laplace transform of the function

$$p(t) = \begin{cases} 1 & \text{when } 0 \leq t < 1 \\ 0 & \text{otherwise} \end{cases}\ .$$

2. Use Table 1 to obtain

(i) $\mathrm{L}^{-1}\left\{\dfrac{(s-3)^2}{s(s^2+9)}\right\}$ (ii) $\left\{\dfrac{1-s-3s^2}{s^2\,(s+1)}\right\}$.

3. Following the working of Example 1.8, show that the Fourier transform of $\dfrac{df}{dt}$ is $jw\,\mathrm{F}(jw)$, assuming that $f(t) \to 0$ as $t \to \pm\infty$.

Answers to Problems

1. $\dfrac{e^{-s}}{s}\ ,\ \dfrac{1-e^{-s}}{s}$.

2. (i) $1 - 2\cos 3t$ (ii) $t + 2 - e^{-t}$.

1.6 TRANSFER FUNCTIONS

In the introduction to this chapter we defined a continuous system as one which can be modelled by an ordinary differential equation. If we restrict our attention to systems described by *linear* differential equations with *constant coefficients*, then the concept of a transfer function becomes useful. For example, consider the system described by the differential equation

$$\frac{d^2y}{dt^2} + 3\frac{dy}{dt} + 2y = 3\frac{dx}{dt} + 2x \qquad (1.13)$$

where $x = x(t)$ and $y = y(t)$ are respectively the input and output of the system. From Problem 3 in Section 1.5 the Fourier transform of $\frac{df}{dt}$ is $jw\ F(jw)$ and extending this result the transform of $\frac{d^2f}{dt^2}$ is $(jw)^2\ F(jw)$. Therefore Fourier transforming equation (1.13) gives

$$(jw)^2\ Y(jw) + 3(jw)\ Y(jw) + 2Y(jw) = 3(jw)\ X(jw) + 2X(jw)$$

where $X(jw)$ and $Y(jw)$ are the Fourier transforms of $x(t)$ and $y(t)$.

Rearranging this transformed equation,

$$\frac{Y(jw)}{X(jw)} = \frac{3(jw) + 2}{(jw)^2 + 3(jw) + 2} = G(jw)\ , \quad \text{say.}$$

$G(jw)$ is called the **transfer function** of the system. Given the input $x(t)$, and assuming that $X(jw)$ exists, $Y(jw)$ can be calculated by simply multiplying by the system transfer function.

$$\text{i.e.} \qquad Y(jw) = G(jw)\ X(jw) \qquad (1.14)$$

$y(t)$ the system output (and the solution of the differential equation) can now be obtained using the inversion formula.

For a system whose input "starts" at $t = 0$ and whose output and its derivatives have zero values at $t= 0$, it is more convenient to use Laplace transforms, since these exist for all normal input functions.

From Example 1.8 $L\left\{\frac{dy}{dt}\right\} = s\ L\{y\}$ when $y(0) = 0$, and this is easily extended to $L\left\{\frac{d^2y}{dt^2}\right\} = s^2L\{y\}$ when $y(0) = \dot{y}(0) = 0$.

Taking Laplace transforms of equation (1.13) leads to

$$L\{y\} = G(s)\ L\{x\} \qquad (1.15)$$

where the *transfer function* $G(s)$ is the same function as $G(jw)$ above with jw replaced by s,

$$\text{i.e.} \qquad G(s) = \frac{3s + 2}{s^2 + 3s + 2}\ .$$

The transfer function of a system relates the *transformed* output and input. $G(s)$ is clearly independent of the input and is a valuable method of representing a linear continuous system.

For a given input function the output of the system can be obtained by taking inverse Laplace transforms of equation (1.15). In Chapters 2 and 3 the applications of transfer functions will be studied in more detail.

Closely related to the idea of a transfer function is the *convolution integral.*

We define the **convolution** of two functions $x(t)$ and $g(t)$ as

$$x(t) * g(t) = \int_0^t x(u)\, g(t-u)\, du$$

where it is assumed that both $x(t)$ and $g(t)$ are zero when $t < 0$.

By considering $x(t)$ as a summation of narrow width delayed pulse functions it is easy to show that the solution of the equation (1.13) is given by

$$y(t) = x(t) * g(t) \tag{1.16}$$

where $g(t)$ is the solution to the differential equation for the special case $x(t) = \delta(t)$, the unit impulse function as defined in Example 1.5. However, we can approach this result by working in the transformed domain (or frequency domain). Following similar working to that given in Example 1.5 we can show that $L\{\delta(t)\} = 1$. For the special case $x(t) = \delta(t)$ equation (1.15) gives $L\{y\} = G(s)$ i.e. $y(t) = L^{-1}\{G(s)\} = g(t)$, the *inverse Laplace transform of the system transfer function*, referred to as the **impulse response function**. For general $x(t)$ the solution is given by equation (1.15) as $y(t) = L^{-1}\{G(s)X(s)\}$, where $X(s)$ denotes $L\{x\}$. To prove equation (1.16) we require that

$$x(t) * g(t) = L^{-1}\{G(s)\, X(s)\} \tag{1.17}$$

Equation (1.17) follows immediately from Theorem 1.1.

Theorem 1.1 (Convolution Theorem)

$$L\{x(t)*g(t)\} = X(s)\, G(s).$$

Proof

$$L\{x(t)*g(t)\} = \int_0^\infty \left\{ \int_0^t x(u)g(t-u)\, du \right\} e^{-st}\, dt$$

$$= \int_0^\infty \int_0^\infty x(u)g(t-u)e^{-st}\, du\, dt, \text{ since } g(t-u) = 0 \text{ for } u > t.$$

Changing the order of integration of the repeated integral we obtain

$$L\{x(t)*g(t)\} = \int_0^\infty x(u) \int_0^\infty g(t-u)e^{-st}\, dt\, du$$

$$= \int_0^\infty x(u) \int_{-u}^\infty g(v)e^{-s(v+u)}\, dv\, du\,, \text{ where } v = t-u$$

$$= \int_0^\infty x(u) \int_0^\infty g(v)e^{-sv}\, e^{-su}\, dv du\,, \text{ since } g(v) = 0 \text{ for } v < 0$$

$$= \int_0^\infty e^{-su}\, x(u)\, du \int_0^\infty e^{-sv}\, g(v)\, dv$$

$$= X(s)\, G(s).$$

Thus we have a choice of methods for solving linear differential equations with constant coefficients.

(i) $y = \int_0^t x(u)\, g(t-u)\, du$, the **time domain** method,

(ii) $y = L^{-1}\{X(s)\, G(s)\}$ the **frequency domain** method.

The following theorem indicates a simple technique for obtaining the impulse response function.

Theorem 1.2 The differential equations

(i) $\ddot{y} + a\dot{y} + by = \delta(y)\ , \quad y(0) = \dot{y}(0) = 0$ and

(ii) $\ddot{y} + a\dot{y} + by = 0, \quad y(0) = 0\ , \ \dot{y}(0) = 1$

have identical solutions.

Proof From Example 1.8 we have

$$L\{\dot{y}\} = s\,L\{y\} - y(0) = s\,L\{y\}$$

Therefore

$$L\{\ddot{y}\} = s\,L\{\dot{y}\} - \dot{y}(0)$$
$$= s^2L\{y\} - sy(0) - \dot{y}(0) = s^2L\{y\} - \dot{y}\,(0)$$

Transforming equation (i) gives

$$s^2L\{y\} + a\,s\,L\{y\} + b\,L\{y\} = 1\ .$$

Therefore

$$L\{y\} = \frac{1}{s^2 + as + b}\ .$$

Transforming equation (ii) gives

$$s^2L\{y\} - 1 + a\,s\,L\{y\} + b\,L\{y\} = 0$$

and

$$L\{y\} = \frac{1}{s^2 + as + b} \quad \text{as in (i).}$$

[This theorem can be generalised to deal with nth order differential equations, where the impulse function on the right-hand side is accommodated by imposing an initial condition on the $(n-1)$th derivative.]

Example 1.9

Solve the differential equation,

$$\ddot{y} + 3\dot{y} + 2y = 2u(t)\ , \quad y(0) = \dot{y}\,(0) = 0\ ,$$

where $u(t)$ is the unit step function. Compare time domain analysis and frequency domain analysis.

(i) **time domain method**

Using Theorem 1.2 the impulse response function $g(t)$ is obtained by solving

$$\ddot{g} + 3\dot{g} + 2g = 0, \ g(0) = 0, \dot{g}(0) = 1 ,$$

i.e. $g(t) = Ae^{-t} + Be^{-2t}$ where $A + B = 0$ and $-A - 2B = 1$.

Therefore $A = 1, B = -1$ and $g(t) = e^{-t} - e^{-2t}$

[Alternatively
$$g(t) = L^{-1}\left\{\frac{1}{s^2 + 3s + 2}\right\}$$
$$= L^{-1}\left\{\frac{1}{s+1} - \frac{1}{s+2}\right\}$$
$$= e^{-t} - e^{-2t} .]$$

Equation (1.16) then gives

$$y(t) = \int_0^t 2(e^{-(t-u)} - e^{-2(t-u)})\,du$$
$$= 2\left[e^{-(t-u)} - \frac{1}{2}e^{-2(t-u)}\right]_0^t$$
$$= 2\left\{\frac{1}{2} - e^{-t} + \frac{1}{2}e^{-2t}\right\}.$$

Therefore the solution is

$$y = 1 - 2e^{-t} + e^{-2t}.$$

(ii) **frequency domain method**

Taking Laplace transforms of the given differential equations

$$(s^2 + 3s + 2)\,L\{y\} = \frac{2}{s} .$$

Therefore
$$L\{y\} = \frac{2}{s(s^2 + 3s + 2)}$$
$$= \frac{1}{s} - \frac{2}{s+1} + \frac{1}{s+2} .$$

Using Table 1 to invert the transforms we obtain

$$y = 1 - 2e^{-t} + e^{-2t} .$$

Strictly the solution should be presented as

$$y = \begin{cases} 1 - 2e^{-t} + e^{-2t} , & \text{for } t \geq 0 \\ 0 , & \text{for } t < 0 \end{cases}$$

since the right-hand side of the given differential equation is zero for $t < 0$.

Alternatively we may write

$$y = (1 - 2e^{-t} + e^{-2t})\,u(t) .$$

Problems

1. Show that $x(t) * g(t) = g(t) * x(t)$.

   Carry out the time domain solution in Example 1.9 using this alternative form of the convolution integral.

2. Solve the differential equation

$$\frac{dy}{dt} + 3y = 13 \cos 2t \ , \quad y(0) = 0 \ ,$$

   using (i) time domain analysis ,

   (ii) frequency domain analysis.

3. Solve equation (1.3) when $x(t) = u(t)$, $y(0) = \dot{y}(0) = 0$.

   Can you explain why your solution does **not** satisfy $\dot{y}(0) = 0$?

Answers to Problems

2. $y = -3e^{-3t} + 3 \cos 2t + 2 \sin 2t$.
3. $y = 1 + e^{-t} - 2e^{-2t}$.

The definition of the impulse function implies that

$$\int_{-\infty}^{t} \delta(u)\, du = \begin{cases} 1 & \text{when } t > 0 \\ 0 & \text{when } t < 0 \end{cases} = u(t) \ ,$$

i.e. $\frac{d}{dt}(u(t)) = \delta(t)$. The equation to be solved is

$$\ddot{y} + 3\dot{y} + 2y = 3\,\delta(t) + 2u(t) \ , \quad y(0) = \dot{y}(0) = 0 \ .$$

Following Theorem 1.2 this is equivalent to

$$\ddot{y} + 3\dot{y} + 2y = 2 \ , \quad y(0) = 0, \ \dot{y}(0) = 3.$$

In fact $\dot{y}(t)$ has a discontinuity at $t = 0$,

i.e. $\dot{y}(0^-) = 0 \ , \quad \dot{y}(0^+) = 3.$

2

Solution Of Linear Differential Equations

2.1 INTRODUCTION

In Example 1.9 we saw how the Laplace transform could be used to solve a certain linear differential equation with constant coefficients. The technique is formalised in this chapter and the concept of a transfer function is again considered. Block diagram notation for continuous systems is introduced, and it is found convenient to describe systems in this form rather than to use the differential equation model. For a given system input we shall see that many properties of the output can be obtained directly from the sytem transfer function without having to solve the corresponding differential equation.

2.2 LAPLACE TRANSFORMS OF COMMON FUNCTIONS

The Laplace transform of a function $f(t)$ is denoted by $F(s)$ and is defined as in equation (1.12),

$$F(s) = \int_0^\infty f(t)\, e^{-st}\, dt \; .$$

An alternative notation uses the *operator* L,

i.e. $$L\{f(t)\} = F(s).$$

The operator transforms a function of t into a function of s, and we saw in Chapter 1 that $s = \sigma + jw$ is a *complex* variable. For all useful functions $f(t)$ the Laplace transform will exist provided that the real part of s is sufficiently large and in general we do not need to worry about convergence of the integral.

For a given function $F(s)$ we can carry out the inverse operation L^{-1}, to obtain $f(t) = L^{-1}\{F(s)\}$, using an inversion formula similar to equation (1.9).

However, for the usual functions it is more convenient to use Table 1 which lists all common transform pairs. In Section 1.5 we obtained the following Laplace transforms

$$L\{u(t)\} = \frac{1}{s} \; ,$$

$$L\{e^{-at}\} = \frac{1}{s+a} \; ,$$

$$L\{\cos wt\} = \frac{s}{s^2+w^2} \; ,$$

$$L\{\sin wt\} = \frac{w}{s^2+w^2} \; .$$

The following examples and theorems extend the list of transforms.

Example 2.1

Obtain the Laplace transform of the function $f(t) = t$.

$$\begin{aligned} L\{t\} &= \int_0^\infty te^{-st} \\ &= \left[\frac{te^{-st}}{-s}\right]_0^\infty + \frac{1}{s}\int_0^\infty e^{-st}\,dt \\ &= 0 - \frac{1}{s^2}\,[e^{-st}]_0^\infty \\ &= \frac{1}{s^2}\ , \quad \text{assuming } \sigma > 0. \end{aligned}$$

Theorem 2.1

Given that $\quad L\{f(t)\} = F(s)$, then

$$L\{tf(t)\} = -\frac{d}{ds}\,F(s)$$

Proof

$$F(s) = \int_0^\infty f(t)\,e^{-st}\,dt$$

Differentiating * with respect to s,

$$\begin{aligned} F'(s) &= -\int_0^\infty tf(t)e^{-st}\,dt \\ &= -L\{tf(t)\} \end{aligned}$$

Therefore $L\{tf(t)\} = -F'(s)$.

Example 2.2

Use Theorem 2.1 to obtain the Laplace transform of

(i) t^2 (ii) te^{-at} .

(i) $$L\{t^2\} = L\{t\,.\,t\} = -\frac{d}{ds}\,L\{t\} = -\frac{d}{ds}\left(\frac{1}{s^2}\right) = \frac{2}{s^3}$$

(ii) $$L\{te^{-at}\} = -\frac{d}{ds}\left(\frac{1}{s+a}\right) = \frac{1}{(s+a)^2}\ .$$

[* This step requires the uniform convergence of the integral

$$\int_0^\infty tf(t)e^{-st}\,dt\ ,$$ see Apostol (1974).]

Theorem 2.2

Given that $L\{f(t)\} = F(s)$, then

$$L\{e^{-at}f(t)\} = F(s+a).$$

Proof

$$L\{e^{-at}f(t)\} = \int_0^\infty e^{-at} f(t)e^{-st}\, dt$$
$$= \int_0^\infty f(t)e^{-(s+a)t}\, dt$$
$$= F(s+a), \text{ i.e. } F(s) \text{ with s replaced by } s+a.$$

Example 2.3

Use Theorem 2.2 to write down the Laplace transforms of $e^{-at}\cos wt$ and $e^{-at}\sin wt$.

$$L\{\cos wt\} = \frac{s}{s^2+w^2} \quad \text{and} \quad L\{\sin wt\} = \frac{w}{s^2+w^2}.$$

Therefore $$L\{e^{-at}\cos wt\} = \frac{s+a}{(s+a)^2+w^2},$$

$$L\{e^{-at}\sin wt\} = \frac{w}{(s+a)^2+w^2}.$$

Example 2.4

Obtain the inverse Laplace transform of $\dfrac{w}{s(s^2+w^2)}$

(i) using partial fractions and Table 1;

(ii) using the convolution theorem (Theorem 1.1)

(i) $$\frac{w}{s(s^2+w^2)} = \frac{1}{w}\left(\frac{1}{s} - \frac{s}{s^2+w^2}\right).$$

Therefore $$L^{-1}\left\{\frac{w}{s(s^2+w^2)}\right\} = \frac{1}{w}(1-\cos t).$$

(ii) From Theorem 1.1

$$L^{-1}\{X(s)\,G(s)\} = \int_0^t x(u)\,g(t-u)\,du.$$

Let $X(s) = \dfrac{1}{s}$, $G(s) = \dfrac{w}{s^2+w^2}$.

Therefore $x(t) = 1$, $g(t) = \sin wt$,

and
$$L^{-1}\left\{\frac{1}{s}\cdot\frac{w}{(s^2+w^2)}\right\} = \int_0^t \sin w(t-u)\,du$$
$$= \left[\frac{1}{w}\cos w(t-u)\right]_0^t$$
$$= \frac{1}{w}(1-\cos wt).$$

Problems

1. Obtain the Laplace transforms of the following functions

(i) t^3 (ii) $e^{-2t}\cos 3t$

(iii) t^2e^{-3t} (iv) $\cosh 2t$ (v) $\cos^2 t$.

2. Find the inverse Laplace transforms of the following functions.

(i) $\dfrac{2}{s(s-1)}$ (ii) $\dfrac{2}{(s+2)^2+4^2}$ (iii) $\dfrac{2}{(s-1)(s^2+1)}$.

3. Apply Theorem 2.1 to show that

$$L\{t^n\} = \frac{n!}{s^{n+1}}, \quad n \text{ a positive integer.}$$

4. (a) Use the convolution theorem to show that

$$L^{-1}\left\{\frac{w}{s^2(s^2+w^2)}\right\} = \frac{wt-\sin wt}{w^2}.$$

(b) Find the inverse Laplace transform of

$$\frac{3s}{(s^2+1)(s^2+4)}.$$

Answers to Problems

1. (i) $\dfrac{6}{s^4}$ (ii) $\dfrac{s+2}{s^2+4s+13}$ (iii) $\dfrac{2}{(s+3)^3}$

(iv) $\dfrac{s}{s^2-4}$ (v) $\dfrac{1}{2}\left(\dfrac{1}{s}+\dfrac{s}{s^2+4}\right)$.

2. (i) $-2+2e^t$ (ii) $\dfrac{1}{2}e^{-2t}\sin 4t$

(iii) $e^t-\cos t-\sin t$.

4. (b) $\cos t - \cos 2t$.

2.3 SOLUTION OF INITIAL VALUE PROBLEMS

An nth order differential equation together with given *initial* values (values at $t = 0$) of the dependent variable and its derivatives up to the $(n-1)$th derivative constitutes an **initial value problem**. For example, the differential equation

$$\ddot{y} + 3\dot{y} + 2y = 1 - t\,,$$

together with the conditions $y(0) = 1$, $\dot{y}(0) = -2$, represents a second order initial value problem.

We restrict our attention to differential equations which are linear with constant coefficients. In this case the application of Laplace transforms results in an *algebraic* equation which can be solved for the Laplace transform of the dependent variable. The method depends on the following two theorems.

Theorem 2.3

$L\{af(t) + bg(t)\} = a\,L\{f(t)\} + b\,L\{g(t)\}$, where a, b are constants.

The proof follows immediately from the definition of the Laplace transform. The theorem is referred to as the **linearity property** of Laplace transforms.

Theorem 2.4

$L\{f^{(n)}(t)\} = s^n L\{f(t)\} - \sum_{r=1}^{n} s^{n-r} f^{(r-1)}(0)$, where $f^{(n)}(t) \equiv \dfrac{d^n f}{dt^n}$ the nth derivative of $f(t)$.

Proof (By induction)

Example 1.8 proved the result for $n = 1$,

i.e. $$L\left\{\frac{df}{dt}\right\} = sL\{f(t)\} - f(0)\,. \qquad (2.1)$$

We assume the result for $n = k$,

i.e. $$L\{f^{(k)}(t)\} = s^k L\{f(t)\} - \sum_{r=1}^{k} s^{k-r} f^{(r-1)}(0)\,. \qquad (2.2)$$

Now $$L\{f^{(k+1)}(t)\} = L\left\{\frac{d}{dt} f^{(k)}(t)\right\}$$

$$= sL\{f^{(k)}(t)\} - f^{(k)}(0)\,,$$

using equation (2.1).

Therefore substituting from equation (2.2)

$$L\{f^{(k+1)}(t)\} = s^{k+1} L\{f(t)\} - \sum_{r=1}^{k} s^{k+1-r} f^{(r-1)}(0) - f^{(k)}(0)$$

$$= s^{k+1} L\{f(t)\} - \sum_{r=1}^{k+1} s^{k+1-r} f^{(r-1)}(0)\,.$$

This last equation represents the result for $n = k + 1$. Therefore we have proved that *if* the result is true for $n = k$, it is *also* true for $n = k + 1$. Since we know that the result is true for $n = 1$, it follows that it is true for $n = 2$, $n = 3$, ..., i.e. for all n.

Examples 2.5, 2.6 and 2.7 demonstrate the application of Laplace transforms to the solution of differential equations.

Example 2.5

Solve the initial value problem

$$\ddot{y} + 3\dot{y} + 2y = 1 - t, \quad y(0) = 1, \quad \dot{y}(0) = -2.$$

From Theorem 2.4

$$L\{\ddot{y}\} = s^2L\{y\} - s\,y(0) - \dot{y}(0) = s^2L\{y\} - s + 2,$$
$$L\{\dot{y}\} = s\,L\{y\} - y(0) = sL\{y\} - 1.$$

The linearity property (Theorem 2.3) allows us to transform the differential equation term by term.

$$s^2\,L\{y\} - s + 2 + 3(sL\{y\} - 1) + 2L\{y\} = \frac{1}{s} - \frac{1}{s^2}$$

i.e. $$(s^2 + 3s + 2)\,L\{y\} = s + 1 + \frac{1}{s} - \frac{1}{s^2} = \frac{s^3 + s^2 + s - 1}{s^2}.$$

Therefore
$$L\{y\} = \frac{s^3 + s^2 + s - 1}{s^2(s^2 + 3s + 2)}$$
$$= \frac{s^3 + s^2 + s - 1}{s^2(s + 1)(s + 2)}$$
$$= \frac{5}{4s} - \frac{1}{2s^2} - \frac{2}{s + 1} + \frac{7}{4(s + 2)}.$$
(partial fractions)

Using Table 1 to invert the Laplace transforms we obtain the required solution,

$$y = \frac{5}{4} - \frac{t}{2} - 2e^{-t} + \frac{7}{4}e^{-2t}.$$

Example 2.6

Solve the differential equation

$$\dddot{y} + 3\ddot{y} + 4\dot{y} + 2y = 10,$$

given the initial conditions $y(0) = \dot{y}(0) = \ddot{y}(0) = 0$.

Since all the initial conditions are zero, Theorem 2.4 gives

$$L\{\dddot{y}\} = s^3L\{y\}, \quad L\{\ddot{y}\} = s^2L\{y\}, \quad L\{\dot{y}\} = sL\{y\}.$$

Transforming the equation gives

$$(s^3 + 3s^2 + 4s + 2)\,\mathrm{L}\{y\} = \frac{10}{s}\ .$$

Therefore
$$\begin{aligned}\mathrm{L}\{y\} &= \frac{10}{s(s^3 + 3s^2 + 4s + 2)}\\ &= \frac{10}{s(s+1)(s^2 + 2s + 2)}\\ &= \frac{5}{s} - \frac{10}{s+1} + \frac{5s}{s^2 + 2s + 2}\\ &= \frac{5}{s} - \frac{10}{s+1} + \frac{5(s+1)}{(s+1)^2 + 1} - \frac{5}{(s+1)^2 + 1}\ .\end{aligned}$$

Note the use of "completing the square" to obtain expressions which appear in Table 1. Inverting the Laplace transforms we obtain the solution,

$$y = 5 - 10e^{-t} + 5e^{-t}\cos t - 5e^{-t}\sin t\ .$$

Example 2.7

Solve the differential equation

$$\ddot{x} - x = \cos 2t\ ,$$

subject to the conditions $x(0) = 0$, $x(\pi) = 0$.

This is not an *initial* value problem since the value of $\dot{x}(0)$ is not given. However, we can assign an arbitrary value A for $\dot{x}(0)$ and subsequently evaluate A to satisfy the condition $x(\pi) = 0$.

Taking Laplace transfroms

$$s^2\,\mathrm{L}\{x\} - \mathrm{A} - \mathrm{L}\{x\} = \frac{s}{s^2 + 4}\ .$$

Therefore $\quad (s^2 - 1)\ \mathrm{L}\{x\} = \mathrm{A} + \dfrac{s}{s^2 + 4}$

i.e.
$$\begin{aligned}\mathrm{L}\{x\} &= \frac{\mathrm{A}}{(s-1)(s+1)} + \frac{s}{(s-1)(s+1)(s^2 + 4)}\\ &= \frac{\mathrm{A}}{2}\left(\frac{1}{(s-1)} - \frac{1}{(s+1)}\right) + \frac{1}{10}\left(\frac{1}{(s-1)} + \frac{1}{(s+1)} - \frac{2s}{(s^2 + 4)}\right).\end{aligned}$$

Therefore $x = \dfrac{\mathrm{A}}{2}\,(e^t - e^{-t}) + \dfrac{1}{10}\,(e^t + e^{-t}) - \dfrac{1}{5}\cos 2t.$

We have to satisfy $x(\pi) = 0$,

i.e. $$0 = \frac{A}{2}(e^{\pi} - e^{-\pi}) + \frac{1}{10}(e^{\pi} + e^{-\pi}) - \frac{1}{5} \ .$$

Therefore $$A = \frac{2 - e^{\pi} - e^{-\pi}}{5(e^{\pi} - e^{-\pi})} \approx -0.1834$$

and the solution is approximately

$$x = 0.0083e^{t} + 0.1917e^{-t} - 0.2\cos 2t \ .$$

Example 2.8

Show that the Laplace transform of $\int_0^t f(u)\,du$ is $\frac{1}{s}$ L$\{f(t)\}$.

Define $$g(t) = \int_0^t f(u)\,du$$ and note that

$$\dot{g}(t) = f(t) \ , \ g(0) = 0.$$

Applying Theorem 2.4,

$$L\{\dot{g}(t)\} = sL\{g(t)\} - g(0)$$

i.e. $$L\{f(t)\} = sL\{g(t)\} \ .$$

Therefore $$L\{g(t)\} = \frac{1}{s}L\{f(t)\}$$, which proves the result.

It is interesting to note that the Laplace transform of an integral involves a division by s, and the Laplace transform of a derivative involves a multiplication by s. This demonstrates a correspondence between the Laplace variable s and the differential operation $D \equiv \frac{d}{dt}$.

Problems

1. Solve the following second order initial value problems:

(i) $\ddot{x} - \dot{x} - 2x = 3e^{2t}$, $x(0) = 0$, $\dot{x}(0) = 2$

(ii) $\ddot{y} + \dot{y} - 2y = \sin t$, $y(0) = 1$, $\dot{y}(0) = -1$

(iii) $\ddot{x} - 2\dot{x} - 3x = 3t^2 + 7t + 3$, $x(0) = \dot{x}(0) = -1$

(iv) $\ddot{y} + 2\dot{y} + 2y = 5\sin t$, $y(0) = \dot{y}(0) = 0$

(v) $\ddot{y} + 2\dot{y} + 2y = 10\sin 2t$, $y(0) = -1$, $\dot{y}(0) = -3$.

2. Solve the following initial value problems:

(i) $\dddot{y} - 3\ddot{y} + 2y = 3e^t$, $y(0) = 0$, $\dot{y}(0) = 1$, $\ddot{y}(0) = 2$

(ii) $\dddot{x} - 2\ddot{x} + \dot{x} = 2$, $x(0) = 2$, $\dot{x}(0) = 1$, $\ddot{x}(0) = -1$

(iii) $\dfrac{d^4\theta}{dt^4} - \theta = 0$, $\theta(0) = 1$, $\dot{\theta}(0) = \ddot{\theta}(0) = \dddot{\theta}(0) = 0$.

3. Use Laplace transforms to find the *general* solution of the following differential equations:

(i) $\dddot{y} - 8y = 16$

(ii) $\ddddot{y} - 2\ddot{y} + 2\dot{y} - y = 0.$

4. Use Laplace transforms to solve the *boundary* value problems:

(i) $\ddot{y} + 0.25y = 0$, $y(0) = 0$, $y(\pi) = 0.5$

(ii) $\ddot{x} + 2\dot{x} = 4t$, $x(0) = x(1) = 0.$

5. Prove that $L\left\{\dfrac{f(t)}{t}\right\} = \int_s^\infty F(u)\,du$, provided that the integral exists, where $F(s) = L\{f(t)\}$. Hence show that $L\left\{\dfrac{\sin t}{t}\right\} = \tan^{-1}\left(\dfrac{1}{s}\right)$.

Answers to Problems

1. (i) $x = \frac{1}{3}(e^{2t} - e^{-t}) + t\,e^{2t}$

(ii) $y = 0.5e^t + 0.6e^{-2t} - 0.1\cos t - 0.3\sin t$

(iii) $x = -1 - t - t^2$

(iv) $y = \sin t - 2\cos t + e^{-t}(\sin t + 2\cos t)$

(v) $y = e^{-t}\cos t - 2\cos 2t - \sin 2t.$

2. (i) $y = \frac{1}{9}(e^t - e^{-2t}) + \frac{2}{3}t\,e^t + \frac{1}{2}t^2e^t$

(ii) $x = 3 + 2t - e^t$

(iii) $\theta = 0.25(e^t + e^{-t}) + 0.5\cos t.$

3. (i) $y = -2 + c_1e^{2t} + e^{-t}(c_2\cos\sqrt{3}\,t + c_3\sin\sqrt{3}\,t)$,

(ii) $y = c_1e^{-t} + (c_2 + c_3t + c_4t^2)e^t$,

where c_1 , c_2 , c_3 , c_4 , are arbitrary constants.

4. (i) $y = 0.5\sin 0.5t$

(ii) $x = -t + t^2.$

2.4 SIMULTANEOUS DIFFERENTIAL EQUATIONS

Continuous systems with multiple inputs and outputs may be represented by systems of simultaneous differential equations. Provided these differential equations are linear with constant coefficients and the inputs belong to a certain class of functions, then they can be expressed in *first order homogeneous form*, $\dot{\underline{x}} = A\underline{x}$ where $\underline{x}$ is an n-dimensional vector and A is an $n \times n$ matrix with constant elements.

Huntley and Johnson (1983) consider a time domain method of solution which depends on algebraic properties of the matrix. Alternatively, the solution may be obtained by taking Laplace transforms of $\dot{\underline{x}} = A\underline{x}$ to obtain

$$sL\{\underline{x}\} - \underline{x}(0) = AL\{\underline{x}\} \qquad (2.3)$$

Equation (2.3) can be arranged to give

$$(sI - A)\, L\{\underline{x}\} = \underline{x}(0)$$

Therefore

$$L\{\underline{x}\} = (sI - A)^{-1}\, \underline{x}(0) \qquad (2.4)$$

and the solution of the system of differential equations is obtained by taking inverse Laplace transforms of equation (2.4). Example 2.9 demonstrates the method. It should be noted that the matrix inverse $(sI - A)^{-1}$ involves division by $|sI - A|$, the determinant of the matrix $sI - A$. If $x = a$ is a solution of the polynomial equation $|sI - A| = 0$ then the solution of the differential equation will include terms e^{at}. The set of solutions of $|sI - A| = 0$ are, of course, the *eigenvalues* of the matrix A, see Huntley and Johnson (1983).

The advantage of Laplace transform techniques applied to simultaneous differential equations is that it is *not necessary* to reduce the equations to the form $\dot{\underline{x}} = A\underline{x}$, see Examples 2.10 and 2.11.

Example 2.9

Solve the system of differential equations

$$\dot{x}_1 = x_1 - 2x_2$$
$$\dot{x}_2 = 5x_1 + 3x_2$$

given the conditions $x_1(0) = 1$, $x_2(0) = 1$.

Writing the system in matrix form

$$\begin{bmatrix} \dot{x}_1 \\ \dot{x}_2 \end{bmatrix} = \begin{bmatrix} 1 & -2 \\ 5 & 3 \end{bmatrix} \begin{bmatrix} x_1 \\ x_2 \end{bmatrix} , \quad \begin{bmatrix} x_1(0) \\ x_2(0) \end{bmatrix} = \begin{bmatrix} 1 \\ 1 \end{bmatrix} .$$

(This step is not strictly necessary; it is easy to transform the equations as they stand and solve for $L\{x_1\}$ and $L\{x_2\}$.

Taking Laplace transforms of the matrix equation gives

$$s \begin{bmatrix} L\{x_1\} \\ L\{x_2\} \end{bmatrix} - \begin{bmatrix} 1 \\ 1 \end{bmatrix} = \begin{bmatrix} 1 & -2 \\ 5 & 3 \end{bmatrix} \begin{bmatrix} L\{x_1\} \\ L\{x_2\} \end{bmatrix} ,$$

i.e. $$\begin{bmatrix} (s-1) & 2 \\ -5 & (s-3) \end{bmatrix} \begin{bmatrix} L\{x_1\} \\ L\{x_2\} \end{bmatrix} = \begin{bmatrix} 1 \\ 1 \end{bmatrix}.$$

Therefore $$\begin{bmatrix} L\{x_1\} \\ L\{x_2\} \end{bmatrix} = \frac{1}{(s^2-4s+13)} \begin{bmatrix} (s-3) & -2 \\ 5 & (s-1) \end{bmatrix} \begin{bmatrix} 1 \\ 1 \end{bmatrix},$$

i.e. $$L\{x_1\} = \frac{s-5}{s^2-4s+13} = \frac{s-2}{(s-2)^2+3^2} - \frac{3}{(s-2)^2+3^2}$$

and $$L\{x_2\} = \frac{s+4}{s^2-4s+13} = \frac{s-2}{(s-2)^2+3^2} + 2\frac{3}{(s-2)^2+3^2}.$$

Inverting the transforms gives the solution

$$x_1 = e^{2t}\cos 3t - e^{2t}\sin 3t$$
$$x_2 = e^{2t}\cos 3t + 2e^{2t}\sin 3t.$$

Example 2.10

Solve the simultaneous differential equations

$$\ddot{x} = 2x + \dot{x} + y$$
$$\dot{y} = 4x + 2y + 1$$

given that $x(0) = 1,\ \dot{x}(0) = 0,\ y(0) = 0$

Taking Laplace transforms of the given equations,

$$s^2L\{x\} - s = 2L\{x\} + sL\{x\} - 1 + L\{y\}$$
$$s\,L\{y\} = 4L\{x\} + 2L\{y\} + \frac{1}{s}$$

i.e. $$(s^2 - s - 2)\,L\{x\} - L\{y\} = s - 1$$
$$-4L\{x\} + (s-2)\,L\{y\} = \frac{1}{s}.$$

Solving these *algebraic* equations for $L\{x\}$ gives

$$L\{x\} = \frac{(s-2)(s-1) + \frac{1}{s}}{(s^2-s-2)(s-2)-4}$$

$$= \frac{s^3 - 3s^2 + 2s + 1}{s^3(s-3)}$$

$$= \frac{20}{27s} - \frac{7}{9s^2} - \frac{1}{3s^3} + \frac{7}{27(s-3)}.$$

Therefore $$x = \frac{20}{27} - \frac{7t}{9} - \frac{t^2}{6} + \frac{7e^{3t}}{27},$$

y can be obtained by solving the algebraic equations for $L\{y\}$, or alternatively by substituting for x in the first differential equation.

i.e. $$y = \ddot{x} - \dot{x} - 2x$$

$$= -\frac{1}{3} + \frac{7e^{3t}}{3} - \left(-\frac{7}{9} - \frac{t}{3} + \frac{7e^{3t}}{9}\right) - \left(\frac{40}{27} - \frac{14t}{9} - \frac{t^2}{3} + \frac{14e^{3t}}{27}\right).$$

Therefore $$y = -\frac{28}{27} + \frac{17t}{9} + \frac{t^2}{3} + \frac{28e^{3t}}{27}.$$

Example 2.11

Consider the undamped mechanical system shown in Figure 2.1. Assuming linear springs and neglecting friction in the wheels the equations of motion are

$$m_1\ddot{y}_1 = k_2(y_2 - y_1) - k_1y_1$$
$$m_2\ddot{y}_2 = -k_2(y_2 - y_1).$$

Solve the equations when $\frac{k_1}{m_1} = 12$,

$\frac{k_2}{m_1} = \frac{k_2}{m_2} = 8$, given the initial conditions

$$y_1(0) = y_2(0) = \dot{y}_1(0) = 0, \quad \dot{y}_2(0) = 24.$$

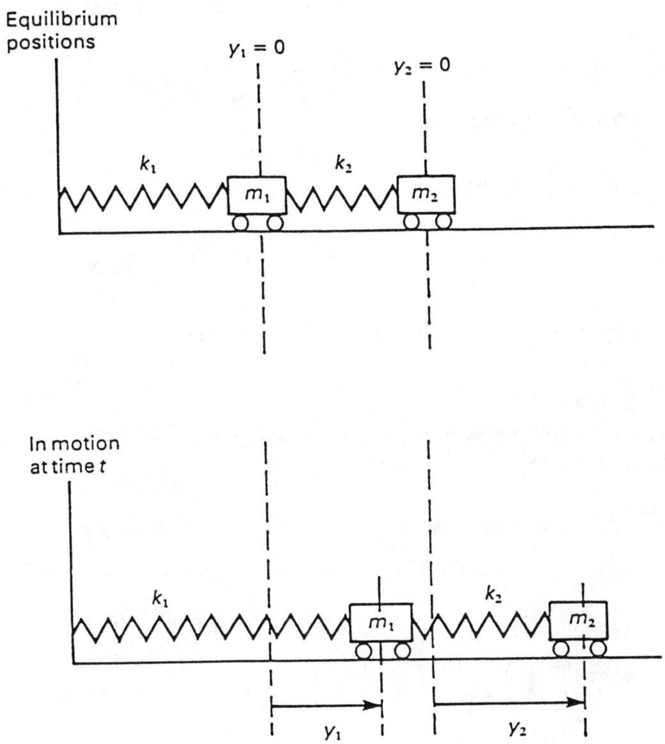

Fig. 2.1 k_1, k_2 are the spring stiffness; m_1, m_2 are the masses of the trucks.

Using the given numbers the equations of motion are

$$\ddot{y}_1 = -20y_1 + 8y_2$$
$$\ddot{y}_2 = 8y_1 - 8y_2 .$$

Taking Laplace transforms we obtain

$$s^2L\{y_1\} = -20\,L\{y_1\} + 8L\{y_2\}$$
$$s^2L\{y_2\} - 24 = 8L\{y_1\} - 8L\{y_2\} ,$$

i.e.

$$(s^2 + 20)\,L\{y_1\} - 8L\{y_2\} = 0$$
$$-8L\{y_1\} + (s^2 + 8)\,L\{y_2\} = 24.$$

Solving for $L\{y_1\}$,

$$L\{y_1\} = \frac{192}{(s^2+20)(s^2+8) - 64}$$
$$= \frac{192}{(s^2+4)(s^2+24)}$$
$$= 4.8\left(\frac{2}{s^2+2^2}\right) - \frac{9.6}{\sqrt{24}}\left(\frac{\sqrt{24}}{s^2+(\sqrt{24})^2}\right) .$$

Therefore $y_1 = 4.8 \sin 2t - 1.96 \sin \sqrt{24}\, t$, approximately.

From the first equation of motion

$$y_2 = \frac{1}{8}(\ddot{y}_1 + 20y_1)$$
$$= \frac{1}{8}(-19.2 \sin 2t + 47.04 \sin \sqrt{24}\, t + 96 \sin 2t - 39.2 \sin \sqrt{24}\, t).$$

Therefore $y_2 = 9.6 \sin 2t + 0.98 \sin \sqrt{24}\, t$, approximately.

Problems

1. Solve the following systems of first order differential equations for the given initial conditions.

(i)
$$\dot{x}_1 = 16x_1 - 8x_2$$
$$\dot{x}_2 = -12x_1 + 12x_2 \quad , \qquad x_1(0) = 2 \ , \ x_2(0) = 3.$$

(ii)
$$\begin{bmatrix} \dot{x}_1 \\ \dot{x}_2 \\ \dot{x}_3 \end{bmatrix} = \begin{bmatrix} 0 & 1 & 1 \\ 0 & -1 & 1 \\ 0 & 0 & -1 \end{bmatrix} \begin{bmatrix} x_1 \\ x_2 \\ x_3 \end{bmatrix} , \qquad \begin{bmatrix} x_1(0) \\ x_2(0) \\ x_3(0) \end{bmatrix} = \begin{bmatrix} 2 \\ 0 \\ -1 \end{bmatrix} .$$

(iii) $\dot{y}_1 = 0.6y_1 + 0.8y_2 - 1$

$\dot{y}_2 = 0.8y_1 - 0.6y_2 + 2$

$y_1(0) = 0\,,\; y_2(0) = 5.$

(iv) $\dot{x} = y + \sin t$

$\dot{y} = 2x + y + 2\cos t$

$x(0) = 0\,,\; y(0) = -1.$

2. Solve the following initial value problems

(i) $\ddot{y} + 6\dot{y} + 8y = 8x$

$\dot{x} + 2x = 0\,,\; x(0) = 2\,,\; y(0) = 1\,,\; \dot{y}(0) = 4.$

(ii) $\dot{x}_1 = x_2$

$\ddot{x}_2 = -\dot{x}_1 + 2\dot{x}_2\,,\; x_1(0) = 1\,,\; x_2(0) = 1\,,\; \dot{x}_2(0) = 2.$

(iii) $\ddot{x}_1 = -10x_1 + 6x_2 + 26e^{-t}$

$\ddot{x}_2 = 15x_1 - 19x_2$

$x_1(0) = 4\,,\; x_2(0) = 3\,,\; \dot{x}_1(0) = 0\,,\; \dot{x}_2(0) = 1.$

3. Figure 2.2 shows a mechanical system in motion at time t ; the positions of the trucks are measured from the equilibrium positions. The equations of motion are

$$m_1\ddot{y}_1 = -(k_1 + k_2)y_1 + k_2y_2$$
$$m_2\ddot{y}_2 = k_2y_1 - (k_2 + k_3)y_2\,.$$

Given that $k_1 = k_2 = k_3 = m_1 = m_2 = 1,$

$y_1(0) = y_2(0) = \dot{y}_1(0) = 0\,,\; \dot{y}_2(0) = 6\,,$

express y_1 and y_2 in terms of t.

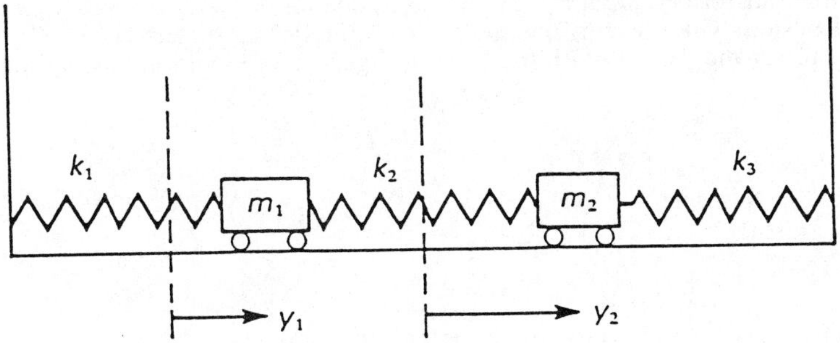

Fig. 2.2 k_1, k_2, k_3 are the spring stiffness; m_1, m_2 are the masses of the trucks.

4. An undamped mechanical system with n degrees of freedom is represented by the system of second order differential equations.

$\ddot{\underline{y}} = A\underline{y}$ where $\underline{y}$ is an n-dimensional vector. Show that, in general, the solution contains terms of frequency $\sqrt{-\lambda}$ where λ is a solution of the polynomial equation $|\lambda I - A| = 0$.

Answers to Problems

1. (i) $x_1 = 2e^{4t}\ ,\ x_2 = 3e^{4t}$

(ii) $$\begin{bmatrix} x_1 \\ x_2 \\ x_3 \end{bmatrix} = \begin{bmatrix} 2+t \\ -t \\ -1 \end{bmatrix} e^{-t}$$

(iii) $y_1 = 2e^{t} - e^{-t} - 1$
$y_2 = e^{t} + 2e^{-t} + 2$

(iv) $x = e^{-t} - \cos t\ ,\ y = -e^{-t}$

2. (i) $x = 2e^{-2t}\ ,\ y = e^{-4t} + 8te^{-2t}$
(ii) $x_1 = 1 + te^{t}\ ,\quad x_2 = e^{t} + te^{t}$

(iii) $x_1 = 2\sin 2t + 4e^{-t}$
$x_2 = 2\sin 2t + 3e^{-t}$

3. $y_1 = 3\sin t - \sqrt{3}\sin\sqrt{3}\,t$
$y_2 = 3\sin t + \sqrt{3}\sin\sqrt{3}\,t\,.$

2.5 BLOCK DIAGRAM NOTATION

We have seen in Section 1.6 that a linear continuous system with a single input and single output (a SISO system) can be represented either by a differential equation or by a transfer function. The differential equation relates the input $x(t)$ to the output $y(t)$, the latter being given by the convolution integral $y(t) = \int_0^t x(u)\, g(t-u)\, du$, where $g(t)$ is the impulse response function. The transfer function $G(s) = L\{g(t)\}$ is a particularly convenient way of representing the system since the transformed output is related to the transformed input by simple multiplication, $L\{y(t)\} = L\{x(t)\}.G(s)$. We refer to the system with transfer function $G(s)$ as simply "the system $G(s)$" and use the convenient **block diagram notation** shown in Figure 2.3 to represent the system.

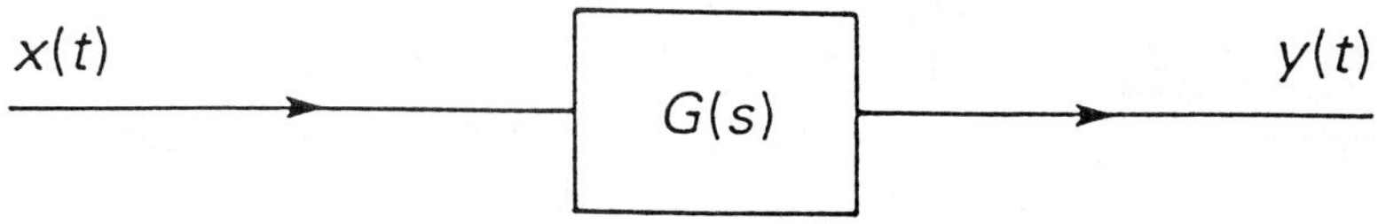

Fig. 2.3 Block diagram representation of system with transfer function G(s), that is, $L\{y(t)\} = L\{x(t)\}\ G(s)$.

It is a simple matter to convert a linear differential equation with constant coefficients to block diagram notation by taking Laplace transforms, assuming that the input, output and their derivatives have zero values when $t = 0$. The reverse process of isolating the differential equation given the block diagram notation is achieved by replacing the Laplace parameter s by the differential operator $D \equiv \frac{d}{dt}$.

Examples 2.12 and 2.13 demonstrate the techniques.

If the output from a system $G_1(s)$ is made the input to a second system $G_2(s)$, then the output of the second system is related to the input of the first system by the transfer function $G_1(s)\ G_2(s)$. In other words the overall transfer function of a number of systems in *series* is obtained by multiplying the individual transfer functions. Similarly, systems in *parallel* can be combined into a single system by addition of the transfer functions, see example 2.14. Figure 2.4 shows the equivalent systems using block diagram notation. Figure 2.5 shows a *feedback system* and its equivalent transfer function. For an introduction to feedback control systems, see Burghes and Graham (1980).

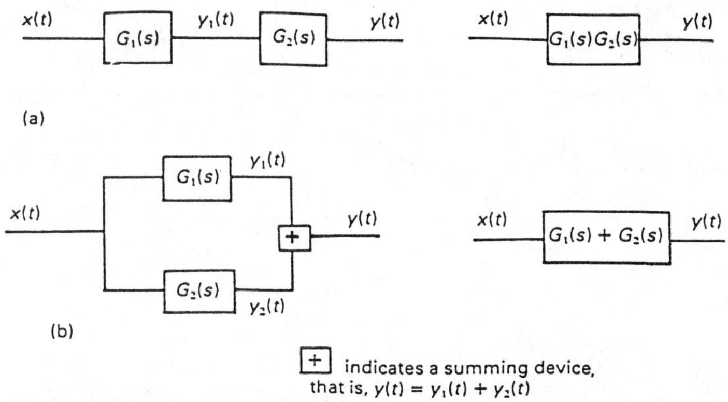

Fig. 2.4 Equivalent systems: (a) systems in series, (b) systems in parallel.

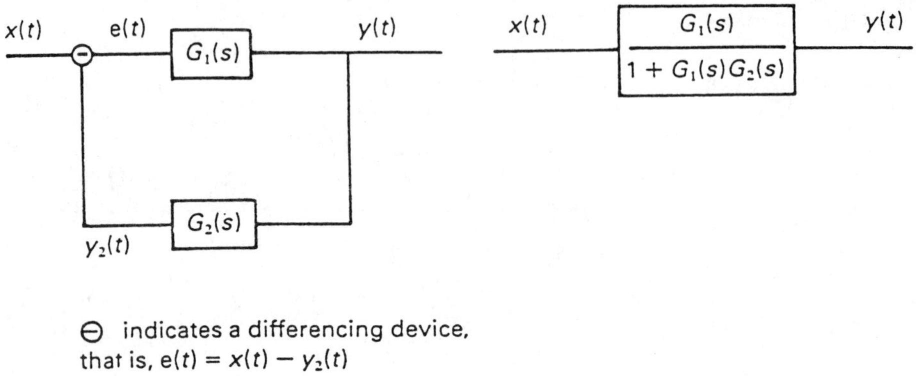

Fig. 2.5 A feedback system and its equivalent.

The idea of transfer function is closely related to concept of *impedance* in a mechanical or electrical system. For electrical systems, taking current as input and voltage as output, the impedance function (or transfer function) for common devices is shown in Figure 2.6, see Holbrook (1966).

Device	Impedance function	Block diagram notation
Resistor R	R	i → [R] → V
Inductor L	Ls	i → [Ls] → V
Capacitor C	$\frac{1}{Cs}$	i → [$\frac{1}{Cs}$] → V

Fig. 2.6 Impedance functions for common devices.

The overall impedance of devices in series is obtained by *adding* their impedances (this is different from transfer functions in series, since a device "output", voltage, is not strictly the "input" to the next device). The derivation of transfer functions for electrical networks is demonstrated in Example 2.15.

Example 2.12

Express the differential equation

$$\ddot{y} + 4\dot{y} + 4y = 4\dot{x} + x$$

in block diagram form. Obtain the output $y(t)$ when the input $x(t)$ is the *unit ramp function,*

$$x(t) = \begin{cases} t & \text{when } t \geq 0 \\ 0 & \text{when } t < 0 \end{cases} .$$

Assuming that $x(0) = y(0) = \dot{y}(0) = 0$ and transforming the differential equation,

$$(s^2 + 4s + 4)\,L\{y\} = (4s + 1)\,L\{x\} .$$

Therefore the transfer function is given by $\dfrac{L\{y\}}{L\{x\}} = \dfrac{4s+1}{s^2+4s+4}$ and in block diagram form the system is

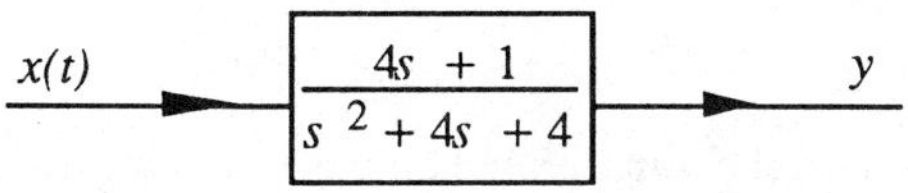

The given input function is the unit ramp function which may be written $x(t) = tu(t)$,

since $$u(t) = \begin{cases} 0 & , \; t < 0 \\ 1 & , \; t \geq 0 \end{cases} .$$

It follows that $L\{x(t)\} = L\{t\,u(t)\} = L\{t\}$, since $u(t) = 1$ for $0 \leq t < \infty$, the range of integration used for calculation of Laplace transforms. In general

$$L\{f(t)\,u(t)\} = L\{f(t)\}.$$

Therefore for the given unit ramp input

$$L\{x(t)\} = \frac{1}{s^2}, \text{ from Table 1, and } L\{y(t)\} = \frac{4s+1}{s^2(s^2+4s+4)}$$

$$= \frac{4s+1}{s^2(s+2)^2}$$

$$= \frac{3}{4s} + \frac{1}{4s^2} - \frac{3}{4(s+2)} - \frac{7}{4(s+2)^2} .$$

Therefore $$y(t) = \frac{3}{4} + \frac{t}{4} - \frac{3e^{-2t}}{4} - \frac{7te^{-2t}}{4} \quad \text{for } t \geq 0.$$

Clearly $y(t) = 0$ for $t < 0$ and we may write the system output in the form

$$y(t) = \left(\frac{3}{4} + \frac{t}{4} - \frac{3e^{-2t}}{4} - \frac{7te^{-2t}}{4}\right) u(t) .$$

Example 2.13

Write down the differential equation which models the system shown in Figure 2.7.

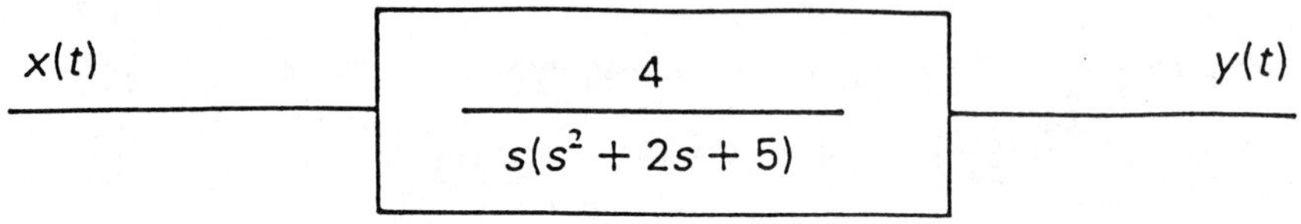

Fig. 2.7

The system is equivalent to the transformed equation

$$L\{y(t)\} = \frac{4}{s(s^2 + 2s + 5)} \cdot L\{x(t)\} .$$

Therefore $\qquad (s^3 + 2s^2 + 5s)\, L\{y(t)\} = 4L\{x(t)\}$

i.e. $\qquad (D^3 + 2D^2 + 5D)\, y(t) = 4x(t)$, where $D \equiv \frac{d}{dt}$.

Therefore the differential equation is

$$\dddot{y} + 2\ddot{y} + 5\dot{y} = 4x(t) .$$

Example 2.14

Prove the equivalence of the systems shown in Figure 2.4.

From Figure 2.4(a) $\qquad L\{y_1(t)\} = L\{x(t)\} \,.\, G_1(s)$

and $\qquad L\{y(t)\} = L\{y_1(t)\} \,.\, G_2(s)$

Therefore $\qquad L\{y(t)\} = L\{x(t)\} \,.\, G_1(s)G_2(s)$

which proves the equivalence.

From Figure 2.4(b) $\qquad L\{y_1(t)\} = L\{x(t)\} \,.\, G_1(s)$

$L\{y_2(t)\} = L\{x(t)\} \,.\, G_2(s)$

and $\qquad L\{y(t)\} = L\{y_1(t)\} + L\{y_2(t)\}$

Therefore $\qquad L\{y(t)\} = L\{x(t)\} \,.\, G_1(s) + L\{x(t)\} \,.\, G_2(s)$

$= L\{x(t)\} \,.\, (G_1(s) + G_2(s))$

which proves the equivalence.

Example 2.15

Obtain the transfer function $\frac{L\{E_0(t)\}}{L\{E_1(t)\}}$ for each of the electrical networks shown in Figure 2.8.

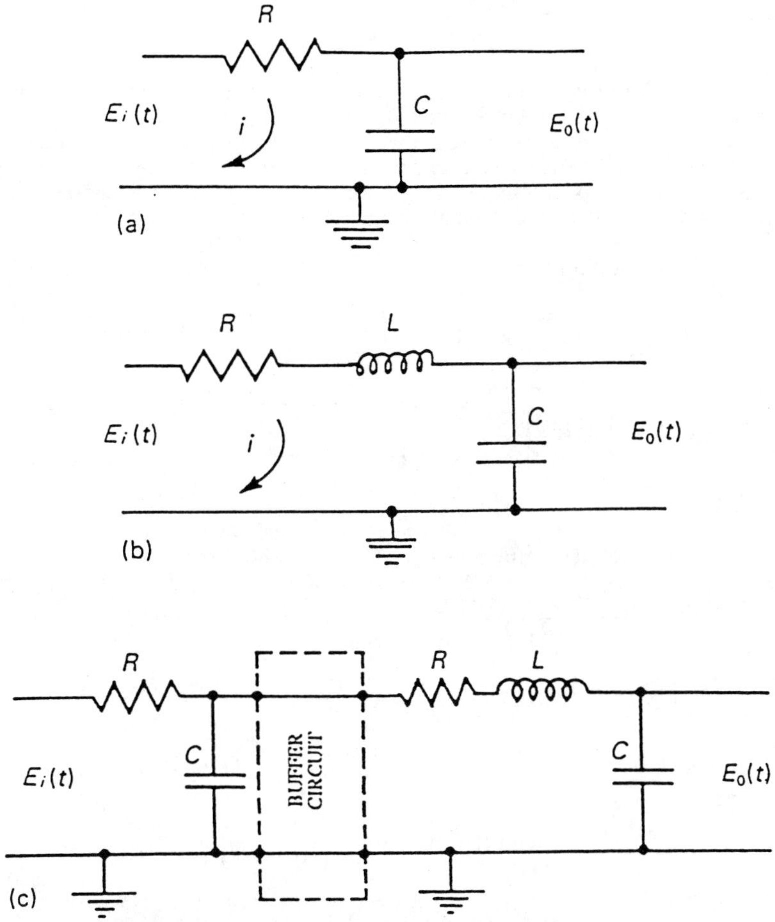

Fig. 2.8 $E_1(t)$, $E_0(t)$ are respectively input and output voltages with reference to earth.

(a) Let i be the current through the capacitor. Adding the impedance functions for the resistor and capacitor we have

$$L\{E_1(t)\} = L(i\}\left(R + \frac{1}{Cs}\right) .$$

Also
$$L\{E_0(t)\} = L\{i\}\frac{1}{Cs} .$$

Therefore
$$\frac{L\{E_0(t)\}}{L\{E_1(t)\}} = \frac{1}{RCs + 1} .$$

(b) Adding the impedance functions for the resistor, inductor and capacitor, we have

$$L\{E_1(t)\} = L\{i\}\left(R + Ls + \frac{1}{Cs}\right)$$

also
$$L\{E_0(t)\} = L\{i\}\,\frac{1}{Cs}\,.$$

Therefore
$$\frac{L\{E_0(t)\}}{L\{E_1(t)\}} = \frac{1}{LCs^2 + RCs + 1}\,.$$

(c) The network in Figure 2.8(c) consists of the two previous networks connected in series through a buffer circuit. The buffer circuit is required to ensure that the output voltage from the first circuit is unchanged when connected to the second circuit. [If the two circuits are connected directly, the ouput voltage of the first circuit as calculated in part (a) is reduced by the presence of the second circuit. This phenomenon is known as *circuit loading*.]

In block diagram notation we have

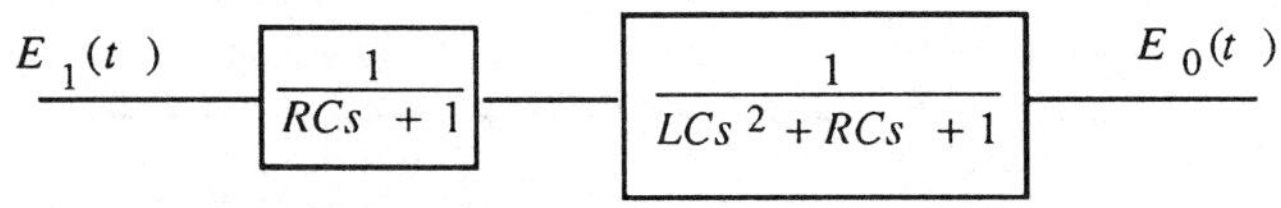

Therefore
$$\frac{L\{E_0(t)\}}{L\{E_1(t)\}} = \frac{1}{(RCs + 1)(LCs^2 + RCs + 1)}\,.$$

Example 2.16

Given that the input voltage in the network of Figure 2.8(a) is given by

$$E_1(t) = \begin{cases} V & \text{when } t \geq 0\,, \quad V \text{ constant} \\ 0 & \text{when } t < 0 \end{cases}$$

find $E_0(t)$. What is the current i at time t ?

From Example 2.15(a)

$$L\{E_0(t)\} = L\{E_1(t)\}\left(\frac{1}{RCs + 1}\right)$$

$$= \frac{V}{s(RCs + 1)}\,, \quad \text{since } E_1(t) = Vu(t)\,,$$

$$= \frac{V}{s} - \frac{V}{s + (RC)^{-1}}\,.$$

Therefore $E_0(t) = V - Ve^{-t/RC}$ for $t \geq 0$, or we may write $E_0(t) = V(1 - e^{-t/RC})\,u(t)$.

We note that $\quad L\{E_0(t)\} = L\{i\}\,\dfrac{1}{Cs}$

i.e. $\quad L\{i\} = Cs\; L\{E_0(t)\}\,.$

This is equivalent to $\quad i(t) = C\,\dfrac{dE_0}{dt}\,.$

Therefore $\quad i(t) = \dfrac{V}{R}\,e^{-t/RC}$ for $t \geq 0$.

Problems

1. Find the outputs of the following systems

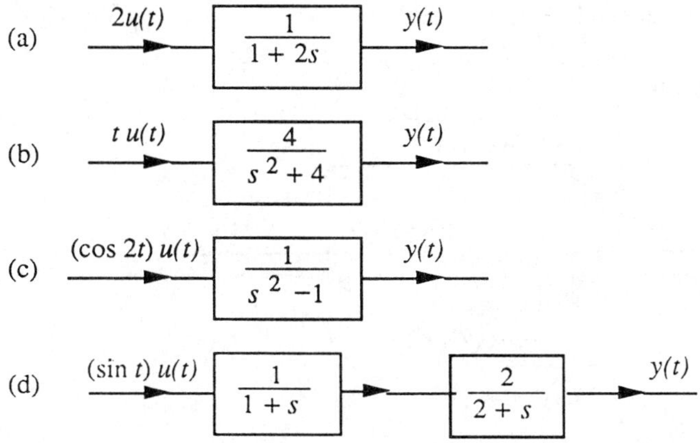

2. Express the following differential equations in block diagram form, and in each case obtain the output $y(t)$ when the input $x(t)$ is a unit step function.

(i) $\ddot{y} + 8y = \dot{x} + 16x$

(ii) $\ddot{y} + 6\dot{y} + 8y = \int_0^t x(u)\, du$.

3. Write down the differential equations described by the diagrams shown in Figure 2.9.

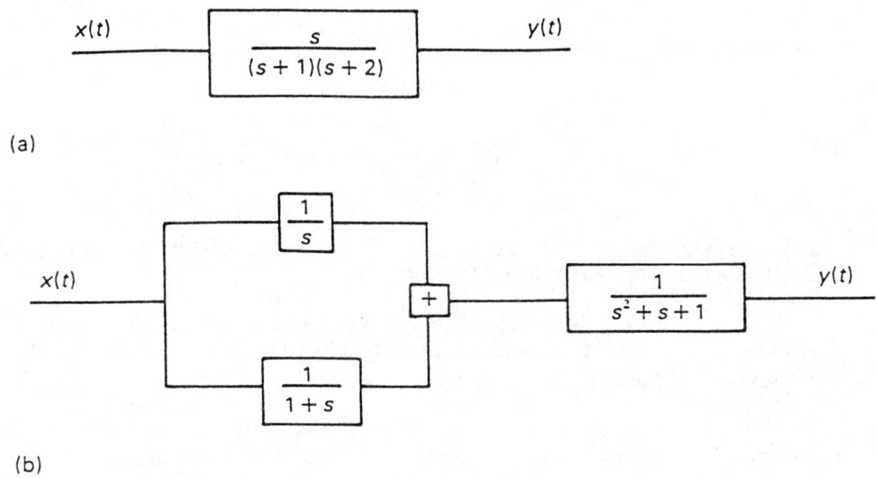

Fig. 2.9

4. Figure 2.10(a) shows a capacitor and a resistor connected in parallel. Assuming that the current $i = i_1 + i_2$ (Kirchoff's law) show that the device has an impedance function, $\frac{R}{RCs + 1}$. Obtain the transfer function $\frac{L\{E_0(t)\}}{L\{E_1(t)\}}$ for the network shown in Figure 2.10(b).

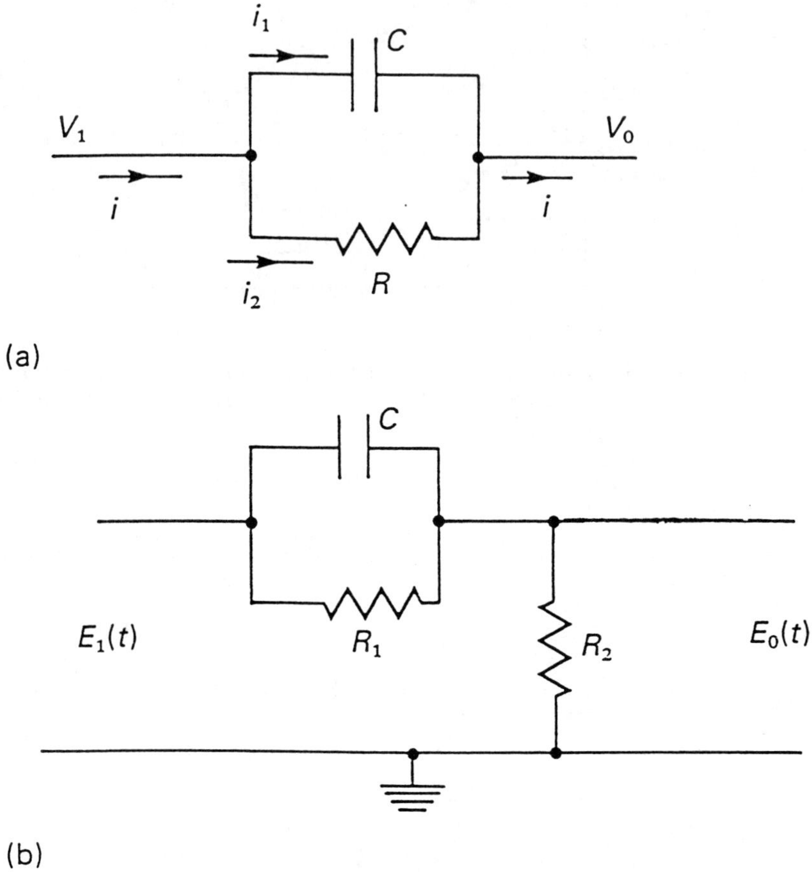

Fig. 2.10

5. Find the output $E_0(t)$ of the network shown in Figure 2.10(b) when the input $E_1(t)$ is a unit step function and $R_1 = R_2 = R$.

6. A constant voltage V is applied at time $t = 0$ across a network consisting of a resistor R, an inductor L and a capacitor C connected in series. Find an expression for the current at time t, given that $R^2C - 4L = 0$.

7. A simple mass-spring system is shown in Figure 2.11(a). Movement of the free end A is considered as the input and the resulting movement of the mass is the output. Neglecting friction the equation of motion of the mass is

$$m\ddot{y} = -k(y - x).$$

Express this equation in block diagram form.

Obtain the transfer function $\frac{L\{y\}}{L\{x\}}$ of the system shown in Figure 2.11(b) and deduce that the double spring system is equivalent to a single spring system of stiffness $(k_1^{-1} + k_2^{-1})^{-1}$. Find the output of the double spring system when x is the unit step function and $y(0) = \dot{y}(0) = 0$.

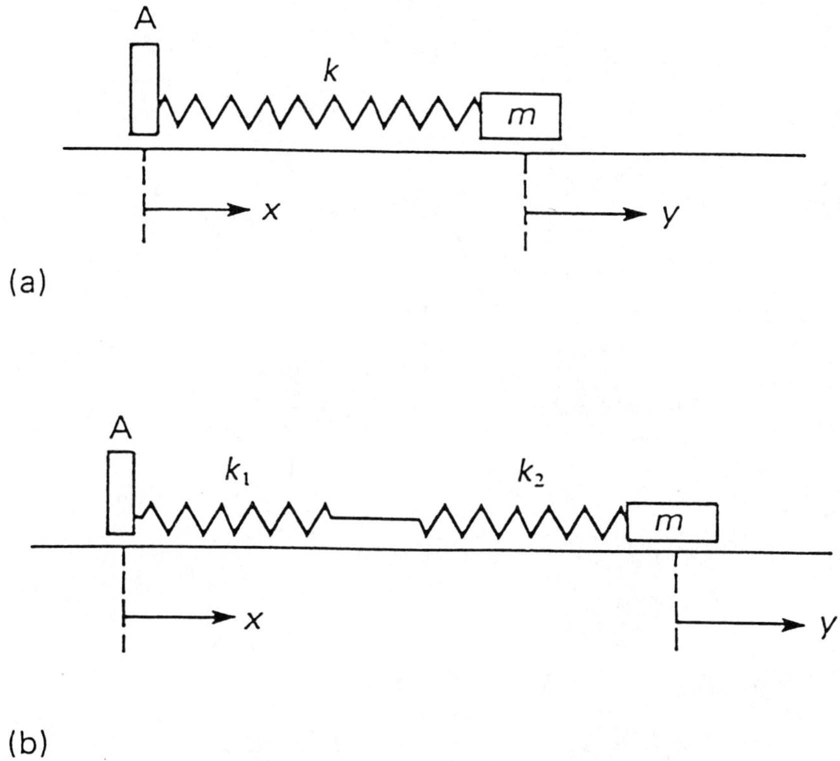

Fig. 2.11 k, k_1, k_2 are the spring stiffnesses.

8. Prove the equivalence of the systems shown in Figure 2.5. Find the output of the feedback system when

$$G_1(s) = \frac{1}{s+3} \quad , \quad G_2(s) = \frac{s+4}{s}$$

and the input is a unit step function.

Answers to problems

1. (a) $2(1 - e^{-0.5t})\, u(t)$

   (b) $\left(t - \frac{1}{2} \sin 2t\right) u(t)$

   (c) $\left[-\frac{1}{5} \cos 2t + \frac{1}{10} (e^{t} + e^{-t})\right] u(t)$

   (d) $\left[\frac{1}{5} (\sin t - 3 \cos t) + e^{-t} - \frac{2}{5} e^{-2t}\right] u(t)$.

2. (i) 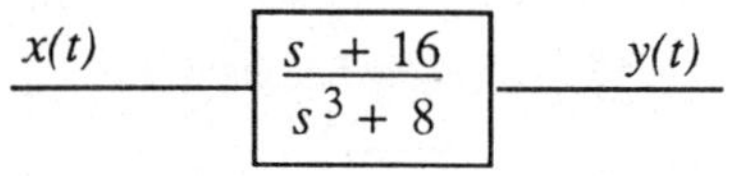

   When $x(t) = u(t)$,

   $$y(t) = \left[2 - \frac{7e^{-2t}}{12} + \frac{e^{t}}{12} (\sqrt{3} \sin \sqrt{3}\, t - 17 \cos \sqrt{3}\, t)\right] u(t) .$$

   (ii) 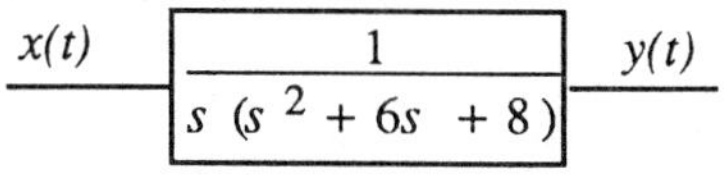

   When $x(t) = u(t)$, $y(t) = \left(\frac{t}{8} - \frac{3}{32} + \frac{e^{-2t}}{8} - \frac{e^{-4t}}{32}\right) u(t)$.

3. (a) $\ddot{y} + 3\dot{y} + 2y = \dot{x}$

   (b) $\dddot{y} + 2\ddot{y} + 2\dot{y} + y = 2\dot{x} + x$.

4. $\frac{R_2(1 + R_1Cs)}{R_1 + R_2 + R_1R_2Cs}$.

5. $E_0(t) = 0.5\,(1 + e^{-2t/RC})\, u(t)$.

6. $i(t) = \frac{Vte^{-Rt/2L}}{L}$, for $t \geq 0$.

7. 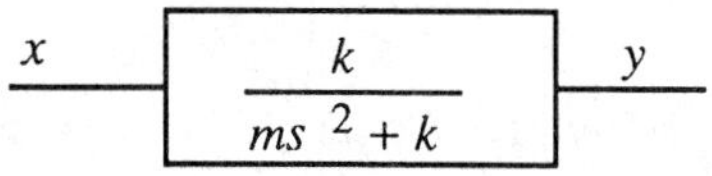

   $y = (1 - \cos wt)\, u(t)$ where $w^2 = \frac{k_1k_2}{m(k_1 + k_2)}$.

8. $y = t\, e^{-2t}\, u(t)$.

2.6 STEADY STATE AND TRANSIENT SOLUTIONS

The output from a system with transfer function G(s) consists of two parts. One part is input independent and contains terms similar to those appearing in the function $g(t) = L^{-1}\{G(s)\}$. The other part is directly related to the input function. In the language of differential equations these parts are referred to respectively as "complementary function" and "particular integral". Control engineers and others using systems notation use the alternative labels "transient response" and "steady state".

The **transient response** of a system is that part of the output which is independent of the input. The **steady state** is the *additional* part of the output which arises due to the presence of the input. The distinction is clearer when dealing with *stable* systems and when the input function does not tend to zero as $t \to \infty$.

[We define a **stable system** as one whose impulse response function $g(t) = L^{-1}\{G(s)\}$ tends to zero as $t \to \infty$.]

For this situation the transient response (as its name suggests) decays to zero as $t \to \infty$ leaving only the steady state output.

For example, consider the system with transfer function $G(s) = \dfrac{2}{(s+1)(s+2)}$: it is a stable system since $g(t) = 2e^{-t} - 2e^{-2t} \to 0$ as $t \to \infty$. If the system input $x(t)$ is a unit step function then $L\{x(t)\} = \dfrac{1}{s}$, $L\{y(t)\} = \dfrac{2}{s(s+1)(s+2)}$ and the system output is given by $y(t) = (1-2e^{-t} + e^{-2t})\,u(t)$. Here the transient response is $y_{TR}(t) = -2e^{-t} + e^{-2t}$ and the steady state is just $y_{ss} = 1$.

Figure 2.12 shows the components of the output for this example.

In Example 2.12 of Section 2.5 we considered $G(s) = \dfrac{4s+1}{s^2+4s+4}$ with $x(t)$ as the unit ramp function. Here $y_{TR} = -\dfrac{3}{4}e^{-2t} - \dfrac{7t\,e^{-2t}}{4}$ and $y_{ss}(t) = \dfrac{3}{4} + \dfrac{t}{4}$.

We have defined the stability of a system in terms of its impulse response function, but it is valuable to investigate what conditions on the system transfer function G(s) are necessary for system stability. We define a **pole** of G(s) as a value of s which makes G(s) infinite.

Assuming that G(s) can be expressed as the quotient of polynomials, then a *pole* of G(s) at $s = \sigma$ is equivalent to a *factor* $(s - \sigma)$ of the denominator polynomial. This in turn gives a term $\dfrac{A}{(s-\sigma)}$ in the partial fraction expansion of G(s) and hence a term $Ae^{\sigma t}$ in the function $g(t)$. Clearly σ must be negative for stability. If G(s) has a pair of complex conjugate poles at $\sigma \pm jw$ then the partial function expansion includes the form $\dfrac{B(s-\sigma) + Cw}{(s-\sigma)^2 + w^2}$ and $g(t)$ contains a term $e^{\sigma t}(B \cos wt + C \sin wt)$. Again $\sigma < 0$ is required for system stability. In general a system will be stable if *all* the poles of the system transfer function G(s) lie in the left hand half of the complex s-plane.

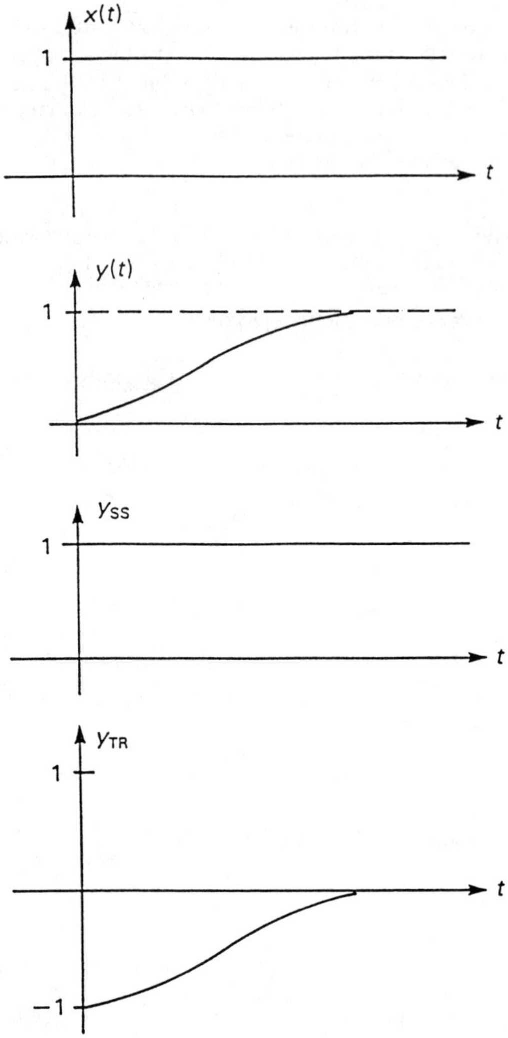

Fig. 2.12 Transient response and steady-state for the system $\frac{2}{(s+1)(s+2)}$.

For systems with simple poles (i.e. non-repeated) on the imaginary axis the function *g(t)* will contain terms such as A, B cos *wt*, C sin *wt* and provided all other poles lie in the left hand half-plane the system is termed **marginally stable.** In all other cases $g(t) \rightarrow \pm\infty$ as $t \rightarrow \infty$ and the system is **unstable**. The **stable region** of the complex *s* plane is shown in Figure 2.13.

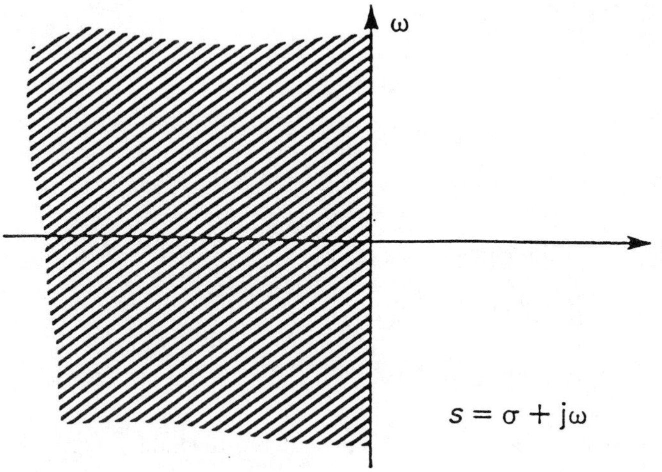

Fig. 2.13 Stable region of the s-plane.

Given that a *stable* system $G(s)$ has an input $x(t) = Ku(t)$, where K is a constant and $u(t)$ is the unit step function, then the steady state output is simply $KG(0)$.

This follows from $L\{y(t)\} = \dfrac{KG(s)}{s}$, which has the term $\dfrac{KG(0)}{s}$ in its partial fraction expansion. This result is a special case of Theorem 2.5.

Theorem 2.5 (final value theorem)

$$\lim_{t\to\infty}\{f(t)\} = \lim_{s\to 0}\{s\,F(s)\}\ , \text{ where }\ F(s) = L\{f(t)\}\ ,$$

assuming the limits exist.

Proof

From Theorem 2.4

$$L\{f'(t)\} = \int_0^\infty f'(t)\,e^{-st}\,dt = s\ F(s) - f(0).$$

Letting $s \to 0$ and assuming the limits exist,

$$\int_0^\infty f'(t)\,dt = \lim_{s\to 0}\{s\,F(s) - f(0)\}$$

Therefore $\qquad f(\infty) - f(0) = \lim_{s\to 0}\{s\,F(s)\} - f(0)\,,$

i.e. $\qquad \lim_{t\to\infty}\{f(t)\} = \lim_{s\to 0}\{s\,F(s)\}.$

Theorem 2.6 (initial value theorem)

$$\lim_{t \to 0^+} \{f(t)\} = \lim_{s \to \infty} \{s\,F(s)\},$$

assuming that the limits exist.

Proof

As in Theorem 2.5

$$\int_0^\infty f'(t)\, e^{-st}\, dt = s\,F(s) - f(0).$$

Now let $s \to \infty$ to obtain, assuming the limits exist,

$$0 = \lim_{s \to \infty} \{s\,F(s) - f(0)\},$$

i.e. $$\lim_{t \to 0^+} \{f(t)\} = \lim_{s \to \infty} \{s\,F(s)\}.$$

Note that we write $t \to 0^+$ since in practice we deal with functions, such as system outputs, which are zero for $t < 0$, i.e. $f(0^-) = 0$.

The initial values and steady state (when it exists) of a system output and its derivatives may be obtained using Theorems 2.5 and 2.6 without the necessity of completing the system solution, see Example 2.19.

For the *special case* when the system input is the *unit step function* the output of the system G(s) is given by

$$L\{y(t)\} = L\{u(t)\}\, G(s) = \frac{1}{s}\, G(s).$$

Applying the Final Value Theorem

$$y_{ss}(t) = \lim_{t \to \infty} \{y(t)\} = \lim_{s \to 0} \left\{s \cdot \frac{1}{s}\, G(s)\right\}$$

$$= G(0) \text{ assuming the limit exists.}$$

Applying the Initial Value Theorem

$$y(0^+) = \lim_{t \to 0^+} \{y(t)\} = \lim_{s \to \infty} \left\{s \cdot \frac{1}{s}\, G(s)\right\}$$

$$= G(\infty) \text{ assuming the limit exists.}$$

Therefore for the unit step input we have the simply calculated results

$$y(\infty) = G(0) \quad , \quad y(0) = G(\infty).$$

[In Chapter 3 we will consider a simple method for obtaining the steady state output of a system when the input is an oscillatory function such as $a \sin wt$ or $a \cos wt$.]

Example 2.17

Find the output of the system with transfer function $G(s) = \dfrac{2s+5}{s^2+2s+5}$ when the input function is

(i) the unit step function,
(ii) the unit ramp function.

In each case separate the steady state and the transient response.

(i) For the unit step function the input is $x(t) = u(t)$.

Therefore $\quad L\{x(t)\} = L\{1\} = \dfrac{1}{s}$ and

$$L\{y(t)\} = \frac{2s+5}{s(s^2+2s+5)}$$

$$= \frac{1}{s} - \frac{s}{s^2+2s+5}$$

$$= \frac{1}{s} - \frac{s+1}{(s+1)^2+2^2} + \frac{\frac{1}{2}(2)}{(s+1)^2+2^2}\,.$$

Therefore $$y(t) = \left(1 - e^{-t}\cos 2t + \frac{1}{2}e^{-t}\sin 2t\right)u(t)\,,$$

i.e. $$y_{ss} = 1\,,\quad y_{TR} = -e^{-t}\left(\cos 2t - \frac{1}{2}\sin 2t\right).$$

(ii) For the unit ramp function the input it $x(t) = t\,u(t)$.

Therefore $\quad L\{x(t)\} = L\{t\} = \dfrac{1}{s^2}$ and

$$L\{y(t)\} = \frac{2s+5}{s^2(s^2+2s+5)}$$

$$= \frac{1}{s^2} - \frac{1}{s^2+2s+5}$$

$$= \frac{1}{s^2} - \frac{\frac{1}{2}(2)}{(s+1)^2+2^2}$$

Therefore $$y(t) = \left(t - \frac{1}{2}e^{-t}\sin 2t\right)u(t)\,,$$

i.e. $$y_{ss} = t\,,\quad y_{TR} = -\frac{1}{2}e^{-t}\sin 2t\,.$$

Example 2.18

For the feedback system shown in Figure 2.14, K is a *positive* constant. Show that the system is stable when the K < 2 and unstable when K > 2.

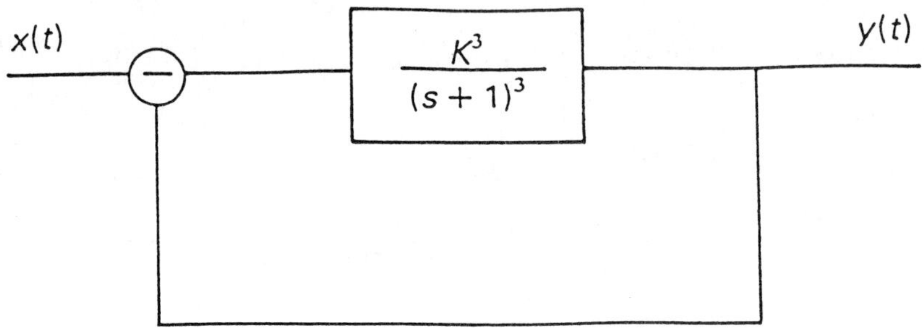

Fig. 2.14

Using the equivalent transfer function shown in Figure 2.5

$$\frac{L\{y(t)\}}{L\{x(t)\}} = \frac{\dfrac{K^3}{(1+s)^3}}{1 + \dfrac{K^3}{(1+s)^3}} ,$$

i.e. the system transfer function is

$$\frac{K^3}{(1+s)^3 + K^3} = \frac{K^3}{(1+s+K)((1+s)^2 - K(1+s) + K^2)}$$

$$= \frac{K^3}{(s+K+1)(s^2 + (2-K)s + K^2 - K + 1)}$$

which has poles at $s = -(K+1)$ and

$$s = \frac{-(2-K) \pm \sqrt{(2-K)^2 - 4(K^2 - K + 1)}}{2} = \frac{-(2-K) \pm j\sqrt{3}\,K}{2} .$$

Therefore when K > 2 there are a pair of complex conjugate poles in the right hand half of the complex s-plane, i.e. the system is unstable. For K < 2 all the poles lie in the left hand half-plane and the system is stable.

Example 2.19

Use Theorems 2.5 and 2.6 to calculate the following properties of the output $y(t)$ for the system with transfer function

$$G(s) = \frac{2s+5}{s^2+2s+5} .$$

(i) $y_{ss}(t)$, $y(0^+)$, $\dot{y}(0^+)$ when the input is a unit step function.

(ii) $\dot{y}_{ss}(t)$, $\ddot{y}(0^+)$ when the input is a unit ramp function.

(i) $$L\{y(t)\} = \frac{2s+5}{s(s^2+2s+5)} \quad \text{for the step input.}$$

$$\begin{aligned} y_{ss}(t) &= \lim_{t\to\infty} y(t) \\ &= \lim_{s\to 0} \left\{\frac{s(2s+5)}{s(s^2+2s+5)}\right\} \\ &= 1. \\ y(0^+) &= \lim_{s\to\infty} \left\{\frac{s(2s+5)}{s(s^2+2s+5)}\right\} \\ &= 0. \end{aligned}$$

Alternatively, since the input is a unit step function,

$$y_{ss}(t) = y(\infty) = G(0) = 1$$

and $$y(0^+) = G(\infty) = 0.$$

$$\begin{aligned} L\{\dot{y}(t)\} &= s\,L\{y(t)\}, \text{ since } y(0^+) = 0, \\ &= \frac{2s+5}{s^2+2s+5} \\ \dot{y}(0^+) &= \lim_{s\to\infty} \left\{\frac{s(2s+5)}{s^2+2s+5}\right\} \\ &= 2. \end{aligned}$$

Note the discontinuity in $\dot{y}(t)$ at $t = 0$.

(ii) $$L\{y(t)\} = \frac{2s+5}{s^2(s^2+2s+5)} \quad \text{for the ramp input.}$$

Therefore $L\{\dot{y}(t)\} = \dfrac{s(2s+5)}{s^2(s^2+2s+5)}$, since $y(0^+) = 0$ for the ramp input,

and $$\begin{aligned} \dot{y}_{ss}(t) &= \lim_{t\to\infty} \dot{y}(t) \\ &= \lim_{s\to 0} \left\{\frac{s^2(2s+5)}{s^2(s^2+2s+5)}\right\} \\ &= 1. \end{aligned}$$

$$\begin{aligned}
L\{\ddot{y}(t)\} &= s^2 L\{y(t)\} \text{ , since } y(0^+) = \dot{y}(0^+) = 0 \quad \text{for the ramp input,}\\
&= \frac{2s+5}{s^2+2s+5}\\
\ddot{y}\,(0^+) &= \lim_{s\to\infty} \left\{\frac{s(2s+5)}{s^2+2s+5}\right\}\\
&= 2.
\end{aligned}$$

[All the answers in this example can be checked from the results of Example 2.17.]

Example 2.20

Calculate $y(0^+)$, $\dot{y}\,(0^+)$ and $y_{ss}(t)$ for the system $G(s) = \frac{s}{1+s}$ with a unit step input. Make a rough sketch of $y \sim t$.

$$\begin{aligned}
y_{ss}(t) &= y(\infty) = G(0) = 0.\\
y(0^+) &= G(\infty) = 1.
\end{aligned}$$

$$\begin{aligned}
L\{\dot{y}\,(t)\} &= s\,L\{y(t)\} - y(0^+)\\
&= s \cdot \frac{1}{s} \cdot \frac{s}{1+s} - 1\\
&= \frac{s}{1+s} - 1\\
&= \frac{-1}{1+s} \;.\\
\dot{y}\,(0^+) &= \lim_{s\to\infty} \left\{\frac{-s}{1+s}\right\} = -1.
\end{aligned}$$

$G(s)$ has one pole at $s = -1$ and $y_{TR}(t)$ will contain only the the term e^{-t} which becomes negligible for $t > 4$.

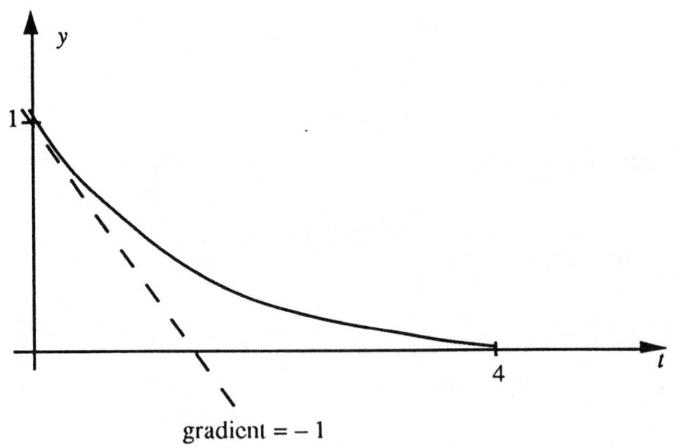

[For this simple example it is easy to obtain the *precise* output $y(t) = e^{-t}\,u(t)$ which confirms the above rough graph.]

Problems

1. Obtain the steady state output and transient response for the following systems with transfer function G(s) and input $x(t)$.

   (i) $G(s) = \dfrac{2}{(1+s)(1+2s)}$, $x(t)$ is unit step function.

   (ii) $G(s) = \dfrac{s+1}{s^2+0.5s+1}$, $x(t)$ is unit ramp function.

   (iii) $G(s) = \dfrac{1}{s^2+2s+1}$, $x(t) = 2 \sin t$.

2. Classify the stability of the following systems.

   (i) $G(s) = \dfrac{1}{s^3+1}$

   (ii) $G(s) = \dfrac{1}{s^2+1}$

   (iii) $G(s) = \dfrac{1}{(s^2+1)^2}$

   (iv) $G(s) = \dfrac{1-s}{s^2+s+1}$

   (v) $G(s) = \dfrac{3s}{s^3+4s^2+5s+2}$

   (vi) $G(s) = \dfrac{s+3}{4s^3+2s^2+s+3}$.

3. The feedback system shown in Figure 2.15 is clearly unstable when $K_2 = 0$. Is it possible to choose a value of K_2 such that the system is stable?

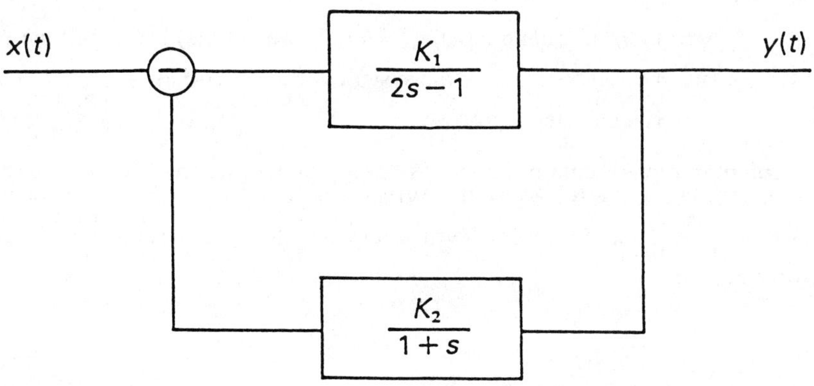

Fig. 2.15

4. For the following systems calculate $y(0^+)$, $\dot{y}(0^+)$, y_{ss} and sketch the graphs of $y \sim t$.

(i)
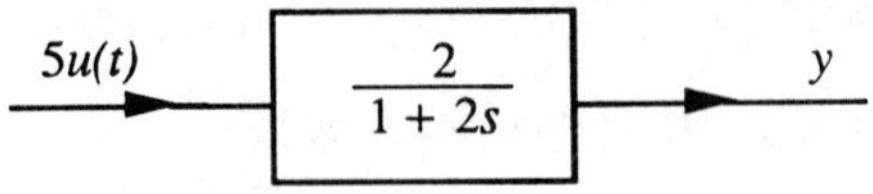

(ii)
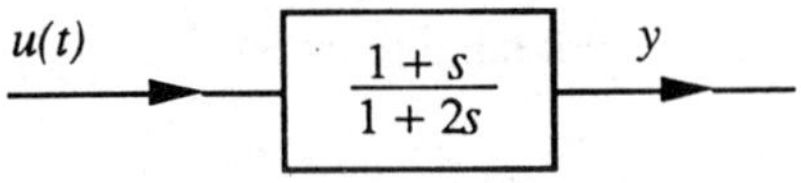

(iii)
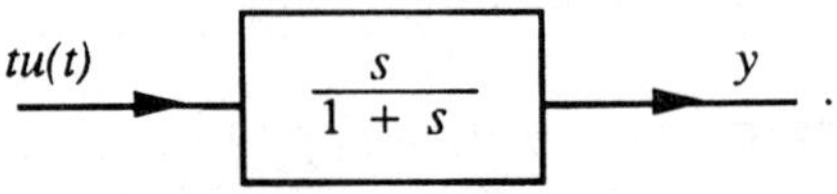

5. Calculate the following without considering the complete solution:

(i) $\dot{y}(0^+)$, $\ddot{y}(0^+)$, $\dot{y}_{ss}$ for system $G(s) = \dfrac{5(s+1)}{s^2 + 0.1s + 2}$

with input $x(t) = t$, $t \geq 0$.

(ii) $y(0^+)$, y_{ss} for system $G(s) = \dfrac{5(s+1)(s+2)}{s^2 + 0.1s + 2}$

with input $x(t) = 3(1 - e^{-3t})$, $t \geq 0$.

6. An important variable in a feedback control system is the error, $e(t) = x(t) - y_2(t)$, which is the difference between input and feedback, see Figure 2.5. Show that $e(t)$ is given by

$$L\{e(t)\} = \frac{L\{x(t)\}}{1 + G_1(s)\,G_2(s)}$$

and write down an expression for the steady state error, e_{ss}.

Given that $G_2(s) = 1$ show that

(i) when $x(t)$ is a step input $e_{ss} = 0$ only when $G_1(s)$ has a pole at $s = 0$,

(ii) when $x(t)$ is a ramp input $e_{ss} = 0$ only when $G_1(s)$ has a multiple pole at $s = 0$. Show that, for the system in figure 2.14, $e_{ss} = \dfrac{1}{1 + K^3}$ when $x(t)$ is a unit step input.

7. A common type of control known as "proportional plus integral" is shown in Figure 2.16, $K_p > 0$, $K_1 > 0$. When $x(t)$ is the unit ramp function and $H(s) = \dfrac{1}{s(s+3)}$, confirm that $e_{ss} = 0$. Show that the system can never be stable when $K_p = 0$.

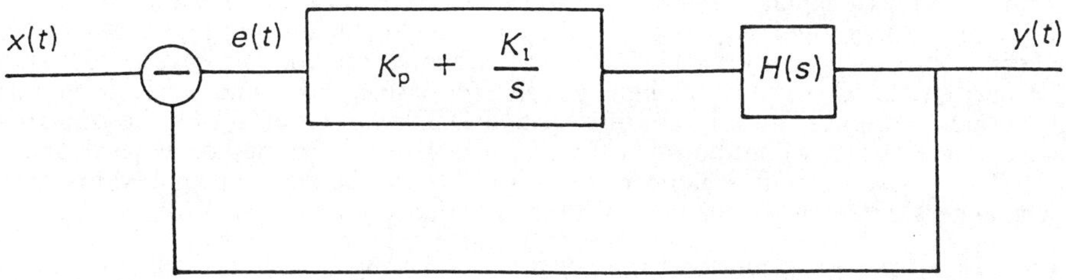

Fig. 2.16 Proportional plus integral control.

Answers to problems

1. (i) $y_{ss} = 2$, $y_{TR} = 2e^{-t} - 4e^{-0.5t}$.

   (ii) $y_{ss} = t + 0.5$, $y_{TR} = -\,e^{-0.25t}\left(\frac{1}{2}\cos kt + \frac{9}{8k}\sin kt\right)$, where $k = \sqrt{\frac{15}{16}}$

   (iii) $y_{ss} = -\cos t$, $y_{TR} = e^{-t}(1 + t)$.

2. (i) Unstable , (ii) marginally stable, (iii) unstable ,

   (iv) stable , (v) stable , (vi) unstable.

3. $K_1K_2 > 1$ for stability.

4. (i) 0 , 5 , 10 (ii) 0.5 , 0.25 , 1 (iii) 0 , 1 , 1 .

5. (i) $\dot{y}(0^+) = 0$, $\ddot{y}(0^+) = 5$, $\dot{y}_{ss} = \frac{5}{2}$.

   (ii) $y(0^+) = 0$, $y_{ss} = 15$.

6. $$e_{ss} = \lim_{s \to 0} \left\{\frac{s\,L\{x(t)\}}{1 + G_1(s)G_2(s)}\right\}.$$

3

Steady State Oscillations

3.1 INTRODUCTION

In the previous chapter we saw that the steady state output of a stable system in response to a step input may be obtained by multiplying the level of the input by G(0) where G(s) is the system transfer function. A similar method is available to obtain the steady state output when the input is a sine (or cosine) wave, and this is developed to define a graphical description of a system which can be used to write down its steady state output when the input frequency is given. The method is particularly useful when dealing with a number of systems in series and may be applied for more general periodic inputs by making use of the Fourier series expansion.

3.2 THE FREQUENCY RESPONSE FUNCTION

When the input to a *linear* system is given by $x(t) = a \sin wt$, (strictly $a \sin wt \,.\, u(t)$), then it is clear that the output will contain terms of frequency w. Further if the system is stable then the steady state output will consist of terms in $\sin wt$ and $\cos wt$ only.

For example consider the system with transfer function

$$G(s) = \frac{1}{1+s} \text{ and input } x(t) = 2 \sin 3t.$$

The output is given by

$$L\{y(t)\} = \frac{6}{(s^2+3^2)(1+s)}$$

$$= \frac{0.6}{s+1} - \left(\frac{0.6s-0.6}{s^2+3^2}\right).$$

Therefore $y(t) = (0.6e^{-t} - 0.6 \cos 3t + 0.2 \sin 3t)u(t)$ and $y_{ss} = -0.6 \cos 3t + 0.2 \sin 3t$.

For the general case $x(t) = a \sin wt$, we note that $a \sin wt$ is the imaginary part of ae^{jwt}, i.e. we may use ae^{jwt} as input and then take the imaginary part of the resulting output.

Thus for a system with transfer function G(s) the transformed output $L\{y(t)\}$ is given by the imaginary part of

$$L\{ae^{jwt}\}\, G(s) = \frac{a\, G(s)}{s-jw}$$

$$= \frac{A}{s-jw} + \frac{B}{s-p_1} + \frac{C}{s-p_2} + \ldots\ldots\ldots$$

where $s = p_1, p_2, \ldots$ are the poles of G(s). Calculations of the constants in the partial fraction expansion gives $A = aG(jw)$ and the system output is given by

$$y(t) = \text{Imaginary Part } \{aG(jw)e^{jwt} + Be^{p_1 t} + Ce^{p_2 t} + \ldots\ldots\}$$

Assuming the system is stable the poles $p_1, p_2, \ldots$ all lie in the left hand half of the complex s-plane, and the steady state output is

$$y_{ss}(t) = \text{Imaginary Part } \{a\text{G}(jw)e^{jwt}\} \qquad (3.1)$$

Now, for a given value of input frequency w, G(jw) is a complex number which may be expressed in polar form

$$\text{G}(jw) = \text{R}e^{j\phi} \qquad (3.2)$$

where R and ϕ depend on w.

Substituting equation (3.2) into (3.1),

$$\begin{aligned} y_{ss}(t) &= \text{Imaginary Part } \{a\text{R}e^{j\phi} e^{jwt}\} \\ &= \text{Imaginary Part } \{a\text{R}e^{j(wt+\phi)}\} \end{aligned}$$

Therefore the steady state output of the system G(s) for input $x(t) = a \sin wt$ is given simply by

$$y_{ss}(t) = a\text{R} \sin (wt + \phi) \qquad (3.3)$$

where R and ϕ are the modulus and argument of the complex number obtained by putting $s = jw$ in the system transfer function. Once again we see the correspondence between the Laplace parameter s and the frequency variable w.

The following points should be noted:

(i) The *frequency* of the output is the same as the input frequency.

(ii) The *amplitude* of the output is obtained by scaling the input amplitude by a factor R. R is called the **gain** of the system G(s), denoted by $|\text{G}(jw))|$.

(iii) There is a *phase difference* of amount ϕ between the output and input, which is equivalent to a time shift of the waveform of amount ϕ/w. ϕ is called the **phase** of the system G(s), denoted by $\underline{|G(jw)}$.

(iv) The steady state output of the system G(s) for input $x(t) = a \cos wt$ is obtained by taking the *real* part in equation (3.1) to give $y_{ss}(t) = a \text{ R} \cos (wt + \phi)$.

The function G(jw) is called the **frequency response function** of the system. It is in fact the *Fourier transform* of the system's impulse response function. Using the frequency response function we are able to write down the steady state oscillatory output whenever the input is a sine (or cosine) wave of given frequency.

Example 3.1

Find the steady state output of the system $\text{G}(s) = \dfrac{1}{1 + 0.1s}$ for the following inputs:

(i) $x(t) = 2 \sin 10t$ (ii) $x(t) = 2 \sin 50t$.

$$G(jw) = \frac{1}{1 + 0.1wj} = Re^{j\phi}$$

where
$$R = \frac{1}{\sqrt{1 + 0.01w^2}}, \quad \phi = -\tan^{-1}(0.1w).$$

(i) When $w = 10$, $R = \frac{1}{\sqrt{2}}$, $\phi = -\frac{\pi}{4}$.

Therefore
$$y_{ss}(t) = 2\left(\frac{1}{\sqrt{2}}\right)\sin\left(10t - \frac{\pi}{4}\right)$$
$$= \sqrt{2}\sin\left(10t - \frac{\pi}{4}\right),$$

i.e. the amplitude has been scaled down by a factor $\frac{1}{\sqrt{2}}$ and there is a **phase lag** of amount $\frac{\pi}{4}$. If we write the output as $y_{ss}(t) = \sqrt{2}\sin 10\left(t - \frac{\pi}{40}\right)$ we see that the output wave "lags" the input wave by an amount $\frac{\pi}{40}$ on the time axis; i.e. the $\frac{\pi}{4}$ phase lag is equivalent to a time axis lag of $\frac{T}{8}$ where $T = \frac{\pi}{5}$ is the period of the input function.

(ii) When $w = 50$, $R = \frac{1}{\sqrt{26}} \approx 0.196$,

$$\phi = -\tan^{-1} 5 \approx -1.373.$$

Therefore $y_{ss}(t) \approx 0.392 \sin(50t - 1.373)$.

For this frequency the amplitude has been scaled down by a factor 0.196 and the phase lag is equivalent to a time lag of 0.0275.

Example 3.2

Obtain the steady state output of the system $G(s) = \frac{1}{s^2 + s + 1}$ with input $x(t) = a\cos wt$, $t \geq 0$. For what value of w is maximum amplification obtained? Write down the output for this case.

$$G(jw) = \frac{1}{1 - w^2 + jw} = Re^{j\phi},$$

where
$$R = \frac{1}{\sqrt{(1 - w^2)^2 + w^2}},$$

$$\text{and} \qquad \phi = \begin{cases} -\tan^{-1}\left(\dfrac{w}{1-w^2}\right) & \text{when } w < 1 \\ -\pi + \tan^{-1}\left(\dfrac{w}{w^2-1}\right) & \text{when } w > 1 \\ -\dfrac{\pi}{2} & \text{when } w = 1 \end{cases}$$

$$\text{and} \qquad y_{ss}(t) = a\,\mathrm{R}\cos(wt + \phi).$$

The amplification factor is R and this is maximised when $(1 - w^2)^2 + w^2$ is minimised.

i.e. when $\dfrac{d}{dw}\left((1 - w^2)^2 + w^2\right) = 0.$

Therefore $-4w + 4w^3 + 2w = 0$ which is satisfied by $w = 0$ and $w = \dfrac{1}{\sqrt{2}}$.

For $w = 0$ the input is a *step function* and the amplification factor is just $G(0) = 1$.

For $w = \dfrac{1}{\sqrt{2}}$ the amplification factor is $\mathrm{R} = \dfrac{1}{\sqrt{0.75}} \approx 1.155$, and $\phi = -\tan(\sqrt{2}) \approx -0.955$.

Thus the maximum amplification is obtained when $w = \dfrac{1}{\sqrt{2}}$ and the corresponding steady state output is $y_{ss}(t) \approx 1.155\, a \cos\left(\dfrac{t}{\sqrt{2}} - 0.955\right)$.

Problems

1. Write down the steady-state outputs of the following systems.

(a)

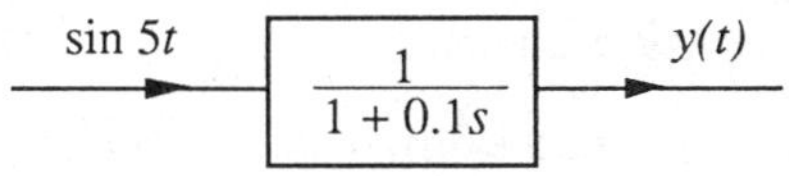

(b)

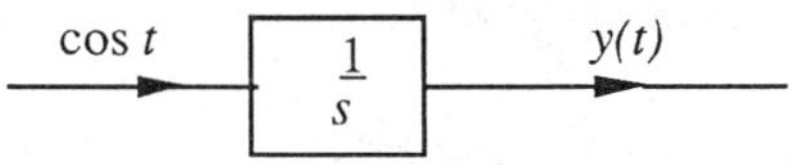

(c)

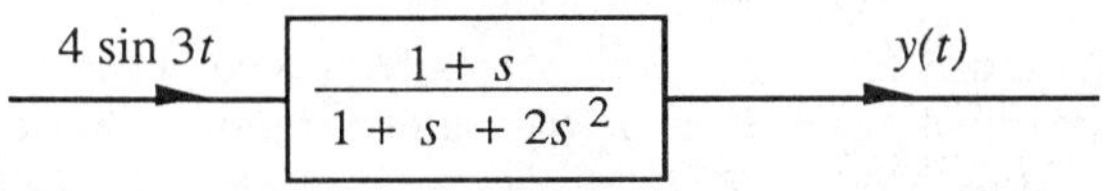

(d)

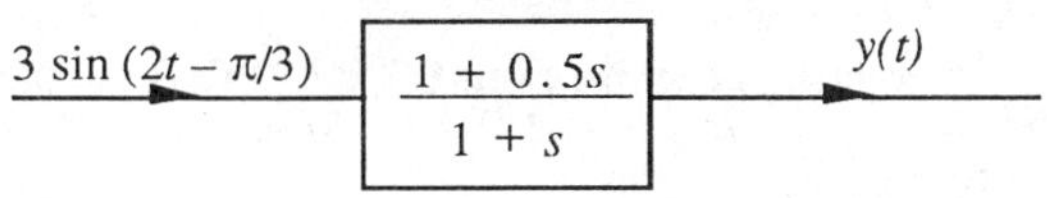

2. Calculate the value of T such that the steady-state output of

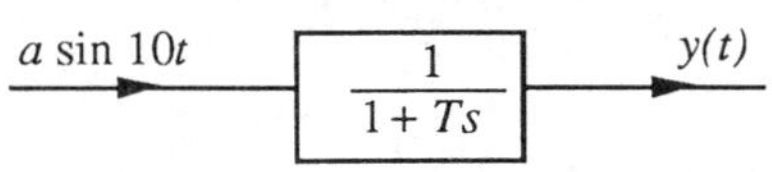

has amplitude $0.5a$.

3. The second order system given by

$$G(s) = \frac{w^2}{s^2 + 2\eta ws + w^2}$$

has input of frequency w. Show that the *gain* of the system is $\frac{1}{2\eta}$ and the *phase* is $-\frac{\pi}{2}$. Write down y_{ss} for

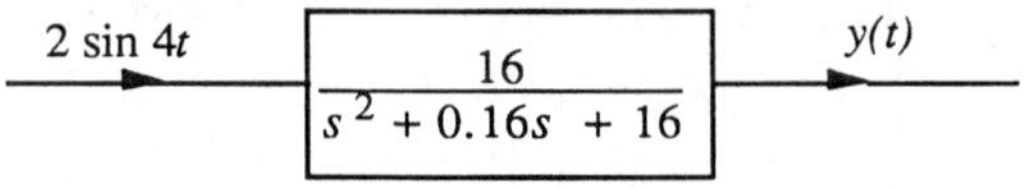

4. Obtain expressions for $|G(jw)|$ and $\angle G(jw)$ for the system $G(s) = \frac{s+1}{s^2 + 4s + 4}$. Write down the steady-state output of the system for each of the following inputs.

(i) $3 \sin 2t$
(ii) $4 \cos 10t$
(iii) $\sin\left(t + \frac{\pi}{3}\right)$.

5. Describe briefly the effect of the system $G(s) = \frac{1+s}{1+2s}$ on the input $a \sin wt$ when

(i) $w << 1$
(ii) $w >> 10$.

At what frequency is the phase lag maximised?

6. Write down the solution for large t of the differential equation

$$\ddot{y} + 2\dot{y} + y = 4 \sin 2t.$$

7. The system $G(s) = \frac{w_n^2}{s^2 + 2\delta_n w_n s + w_n^2}$, $0 < \delta_n < \frac{1}{\sqrt{2}}$, has input $x(t) = a \sin wt$ (w_n and δ_n are constants known respectively as natural frequency and damping factor). For what value of w is the output amplitude maximised? For this frequency find the value of δ_n to give an amplification factor of 10.

8. For the electrical network shown in Figure 2.10(b) find the steady state output voltage E_0 when $E_1 = 20 \sin 3t$ and $R_1C = R_2C = 0.2$. What is the steady output when the input frequency is

(i) very low
(ii) very high?

Answers to Problems

1. (a) $0.894 \sin (5t - 0.464)$ (b) $\cos\left(t - \frac{\pi}{2}\right) = \sin t$

   (c) $0.733 \sin (3t - 1.718)$ (d) $1.897 \sin (2t - 1.369)$

2. 0.1732

3. $50 \sin\left(4t - \frac{\pi}{2}\right) = - 50 \cos 4t.$

4. $|G(jw)| = \sqrt{\frac{1 + w^2}{(4 - w^2)^2 + 16w^2}}$

$$\underline{/G(jw)} = \begin{cases} \tan^{-1} w - \tan^{-1}\left(\frac{4w}{4 - w^2}\right) & \text{when } w < 2 \\ \tan^{-1} w - \pi + \tan^{-1}\left(\frac{4w}{w^2 - 4}\right) & \text{when } w > 2 \\ \tan^{-1} 2 - \frac{\pi}{2} & \text{when } w = 2 \end{cases}$$

   (i) $-\frac{3\sqrt{5}}{8} \cos (2t + \gamma)$ where $\gamma = \tan^{-1} 2$.

   (ii) $-\sqrt{\frac{101}{676}} \cos (10t + \beta)$ where $\beta = \tan^{-1} 10 + \tan^{-1}\left(\frac{5}{12}\right)$.

   (iii) $\frac{\sqrt{2}}{5} \sin (t + \alpha)$ where $\alpha = \frac{7\pi}{12} - \tan^{-1}\left(\frac{4}{3}\right)$.

5. (i) output $\approx$ input for small w.

   (ii) output $\approx \frac{1}{2}$ (input) for large w

   $w = \frac{1}{\sqrt{2}}$ to maximise phase lag.

6. $y_{ss} = - 0.8 \sin (2t + \alpha)$ where $\alpha = \tan^{-1}\left(\frac{4}{3}\right)$

   or $y_{ss} = - 0.48 \sin 2t - 0.64 \cos 2t$..

7. $w = w_n \sqrt{1 - 2\delta_n^2}$, $\delta_n \approx 0.05$.

8. $E_{0_{ss}} \approx 11.17 \sin (3t + 0.249)$

   (i) $E_{0_{ss}} \approx 10 \sin wt$ when w small

   (ii) $E_{0_{ss}} \approx 20 \sin wt$ when w large.

3.3 BODE DIAGRAMS

The graphs of $|G(jw)|$ and $\angle G(jw)$ against w are called the **Bode diagrams** for the system $G(s)$. Together the two graphs are an alternative representation of the system since the functions $|G(jw)|$ and $\angle G(jw)$ completely define the system transfer function $G(jw)$. Given the Bode diagrams for a system and the input $x(t) = a \sin wt$, the steady state output $y_{ss}(t) = aR \sin (wt + \phi)$ can be written down by simply reading off from the graphs $R = |G(jw)|$ and $\phi = \angle G(jw)$. The Bode diagrams for the first order system $G(s) = \dfrac{1}{1 + sT}$ are shown in Figure 3.1, using the *non-dimensional form* wT on the horizontal axis. It is convenient to plot wT on a logarithmic scale and to present $|G(jw)|$ in the dB logarithmic form*, defined by

$$R\,dB = 20 \log_{10} R$$

The reason for this log-log presentation is partly because the graph of $|G(jw)|$ becomes a straight line for large w, but more importantly because the *gains* of two systems in series may be *added* when in dB form. This follows because for systems $G_1(jw) = R_1 e^{j\phi}$ and $G_2(jw) = R_2 e^{j\phi_2}$ in series the equivalent system $G(jw) = Re^{j\phi}$ is given by $G(jw) = G_1(jw)\,G_2(jw) = R_1R_2\, e^{j(\phi_1 + \phi_2)}$, i.e. $\phi = \phi_1 + \phi_2$ and $R = R_1R_2$, but in dB form $R dB = R_1 dB + R_2 dB$. Therefore both the gain R *and* the phase ϕ of the equivalent system $G(jw)$ can be obtained by addition.

The Bode diagrams for the second order system $G(s) = \dfrac{w_n^2}{s^2 + 2\delta_n w_n s + w_n^2}$ are shown in Figure 3.2 for a range of value of δ_n; the non-dimensional form w/w_n is used on the horizontal axis. w_n and δ_n are called the **natural frequency** and **damping factor** of the system.

Usually the transfer function of a system can be expressed as the quotient of polynomials and, since polynomials can be factorised into linear and quadratic factors, Figures 3.1 and 3.2 may be used to obtain the steady state oscillatory output of the system, see Example 3.4. In certain circumstances the transfer function of a particular system may be unknown but, assuming the system is linear, it is still possible to obtain the Bode diagrams by experiment using a transfer function analyser. This equipment generates a sine wave input and measures the amplification and phase shift of the output for a range of input frequencies. The resulting Bode diagrams may give some information about the system's transfer function. For example, the slope of the $|G(jw|$ plot for large w is related to the difference in degree of the numerator and denominator polynomials of $G(s)$. However, if the system is just one component of a large system, then the steady state oscillations of the latter may be determined by consideration of Bode diagrams: it is not necessary to know the transfer functions of every component.

* dB denotes *decibels*, units used by systems engineers to measure amplitude of waves. Note that $R = 2$ is equivalent to $R\,dB \approx 6$ and $R = \frac{1}{2}$ is equivalent to $R dB \approx -6$.

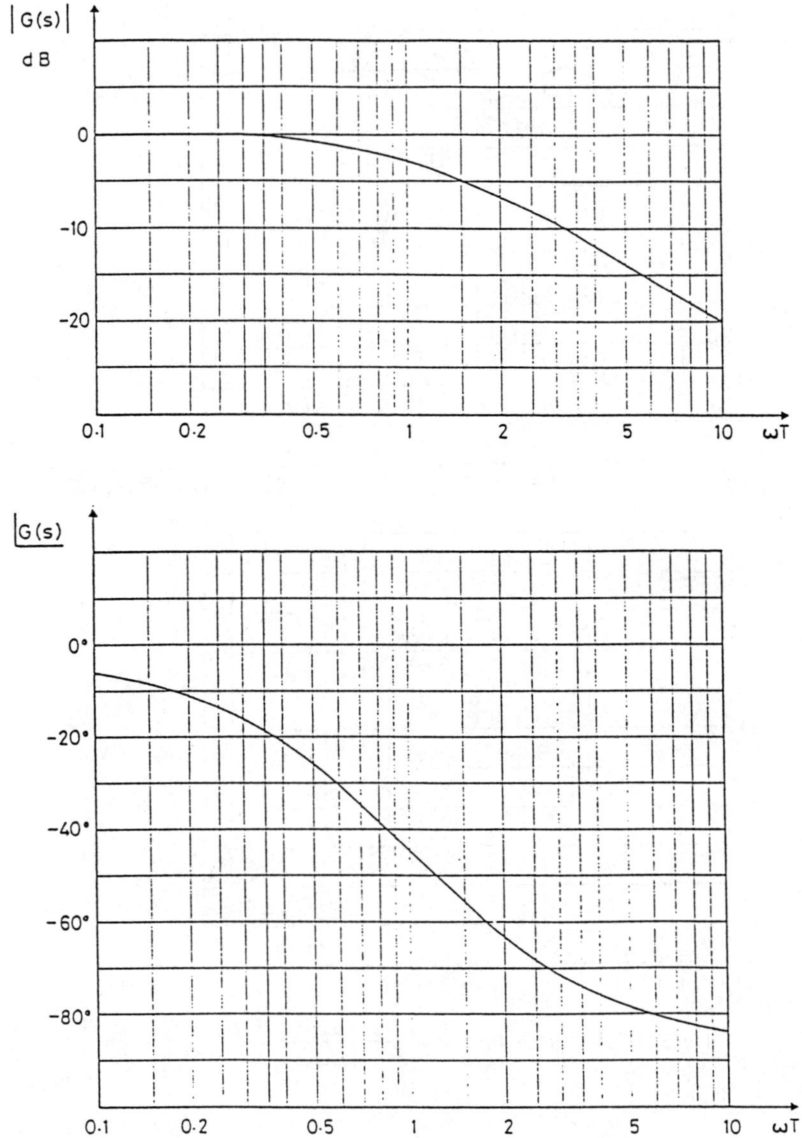

Fig. 3.1 Bode diagram for $G(s) = \dfrac{1}{1 + sT}$.

Example 3.3

Use Figure 3.1 to obtain the steady state output of the system $G(s) = \dfrac{1+s}{1+2s}$ when the input is $x(t) = 5 \sin 2t$.

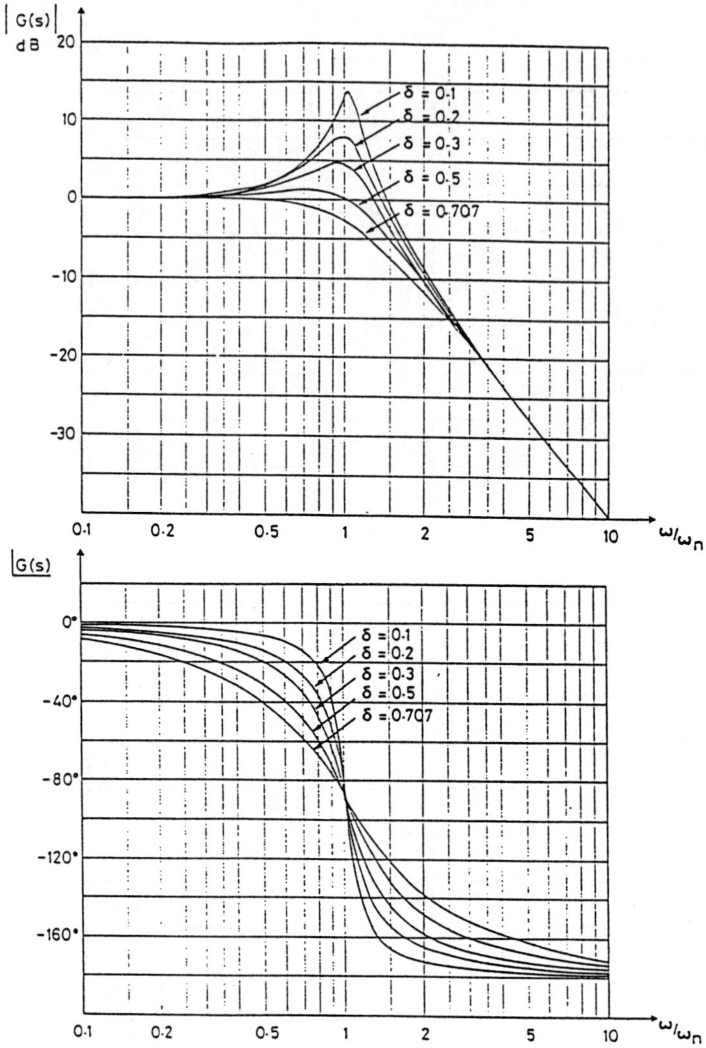

Fig. 3.2 Bode diagram for $G(s) = \dfrac{\omega_n^2}{s^2 + 2\delta_n\omega_n s + \omega_n^2}$.

The system G(s) can be considered as two systems $G_1(s) = \dfrac{1}{1+2s}$ and $G_2(s) = 1 + s$ in series. The gain and phase of $G_1(s)$ can be read directly from Figure 3.1 for $w\mathrm{T} = 2 \times 2 = 4$. The gain and phase of $G_2(s)$ are obtained by applying Figure 3.1 to the function $\dfrac{1}{1+s}$, $w\mathrm{T} = 2 \times 1 = 2$, and *changing the sign* of both R dB and ϕ, since $\dfrac{1}{G_2(s)} = \dfrac{1}{R_2 e^{j\phi_2}} = \dfrac{1}{R_2} e^{-j\phi_2}$ and $\dfrac{1}{R_2}\,d\mathrm{B} = -R_2 d\mathrm{B}$.

Therefore $|G_1(j\,2)| = -12.3dB$ $\quad \angle G_1(j\,2) = -76°$

and $|G_2(j\,2)| = 7dB$ $\quad \angle G_2(j\,2) = 63.5°$.

Therefore $|G(j\,2)| = -12.3 + 7 = -5.3dB$

and $\angle G(j\,2) = -76° + 63.5° = -12.5°$.

It follows that the gain factor R = 0.543 and the phase shift $\phi = -0.218$ (radians), i.e. the steady state output is

$$y_{ss}(t) = 2.71 \sin(2t - 0.218) \text{ approximately.}$$

Example 3.4

Use Figures 3.1 and 3.2 to obtain the steady state output of the system

$$G(s) = \frac{50(s+20)}{s^3 + 14s^2 + 140s + 1000}$$

when the input is $x(t) = \cos 20t,\ t \geq 0$.

Factorising the denominator of G(s) gives

$$G(s) = \frac{50(s+20)}{(s+10)(s^2+4s+100)} = \frac{100(1+0.05s)}{(1+0.1s)(s^2+4s+100)} \ .$$

We can consider G(s) as equivalent to the following three systems in the series.

$$G_1(s) = 1 + 0.05s, \quad \text{i.e.} \quad T = 0.05,$$

$$G_2(s) = \frac{1}{1+0.1s} \quad \text{i.e.} \quad T = 0.1,$$

$$G_3(s) = \frac{100}{s^2+4s+100} \quad \text{i.e.} \quad w_n = 10\ ,\ \delta_n = 0.2.$$

For the given input function the frequency $w = 20$, and from Figure 3.1

$|G_1(j\,20)| = 3\ dB$ $\quad \angle G_1(j\,20) = 45°$

$|G_2(j\,20)| = -7\ dB$ $\quad \angle G_2(j\,20) = -63.5°$

From Figure 3.2

$|G_3(j\,20)| = -9.8\ dB$ $\quad \angle G_3(j\,20) = -165.1°$

Therefore since $G(s) = G_1(s)\,.\,G_2(s)\,.\,G_3(s)$.

$|G(j\,20)| = 3 - 7 - 9.8 = -13.8\ dB$ (i.e. R = 0.2)

$\angle G(j\,20) = 45° - 63.5° - 165.1° = -183.6°$ (i.e. $\phi = -3.20$)

and the steady state output is

$$y_{ss}(t) = 0.2 \cos(20t - 3.2) \text{ approximately.}$$

Example 3.5

A system G(s) consists of two subsystems $G_1(s)$ and $G_2(s)$ connected in series. $G_1(s) = \frac{1}{1 + 0.02s}$ but $G_2(s)$ is unknown. However, Bode diagrams for the second subsystem were obtained by experiment and the following table contains particular readings from these graphs.

Frequency w rads/sec	$\lvert G_2(jw) \rvert$	$\angle G_2(jw)$
0	12 dB	0°
10	13 dB	–34°
150	– 23 dB	– 172°

Find the output of the system G(s) for the input $x(t) = 2 + \sin 10t + 0.5 \sin 150t$.

Using Figure 3.1 we can construct the corresponding table for $G_1(s)$ and hence G(s) since $G(s) = G_1(s) . G_2(s)$.

w	$\lvert G_1(jw) \rvert$	$\angle G_1(jw)$
0	0 dB	0°
10	– 0.2 dB	– 11.3°
150	– 10dB	– 71.6°

w	$\lvert G(jw) \rvert$	$\angle G(jw)$
0	12 dB	0°
10	12.8 dB	–45.3°
150	– 33 dB	– 243.6°

The corresponding amplification factors and phase shifts for the system G(s) are given below.

w	R	ϕ (radians)
0	4	0
10	4.4	– 0.79
150	0.02	– 4.25

Therefore the steady state output of the systems G(s) is

$$y_{ss}(t) = 8 + 4.4 \sin (10t - 0.79) + 0.01 \sin (150t - 4.25).$$

(The linearity property allows us to add the outputs from each component of the input.)

Problems

1. Use Figures 3.1 and 3.2 to confirm the results of problems 4, 6 and 8 in Section 3.2.

2. Apply Figures 3.1 and 3.2 to obtain the steady state output of the system

$$G(s) = \frac{s^2 + 16s + 100}{(s+4)(s^2+4s+16)}$$

when the input function is $x(t) = 1 - 2\cos 6t$.

3. Solve the differential equation

$$\ddot{y} + 3\dot{y} + 2y = 3\dot{x} + 2x.$$

where $x = \cos 2t$ and $y(0) = \dot{y}(0) = 0$. Use Figure 3.1 to confirm the steady state solution.

4. Control engineers often approximate to $|G(jw)|$ for $G(s) = \dfrac{1}{1+sT}$ as follows,

$$|G(jw)| = \begin{cases} 0\ dB & \text{for } wT < 1 \\ -20\log_{10}(wT)\ dB & \text{for } wT > 1 \end{cases}.$$

Show that the maximum error in the gain factor R caused by this approximation is just above 40%. At what frequency does the maximum error occur?

5. The guidance system of an air-to-air missile measures $\dot{\phi}$ the rate of change of the line of sight to target. The output from the guidance system is a voltage V which is input to the missile control system to produce a lateral acceleration $\ddot{y}$.

An approximation to the guidance system is $\dfrac{L\{V\}}{L\{\dot{\phi}\}} = \dfrac{2}{1+0.1s}$ volts/rad sec^{-1},

and the missile control system is $\dfrac{L\{\ddot{y}\}}{L\{V\}} = G(s)$, where $G(jw)$ has been determined by experiment in the form of Bode diagrams.

In particular $G(0) = 300$ ms^{-2} / volt, $\left|\dfrac{G(j5)}{G(0)}\right| = 2dB$, $\underline{/G(j5)} = -40°$. Given that the target manoeuvres in such a way that $\dot{\phi} = 0.1(1 - \cos 5t)$, find the steady-state lateral acceleration of the missile.

6. The pressure of liquid in a pipeline is varying sinusoidally (i.e. as a sine wave) between 1500 psi and 3000 psi at a frequency of 10 Hz (i.e. $w = 20\pi$). A pressure gauage connected to the pipeline may be considered as a first order system $G(s) = \dfrac{1}{1+0.2s}$. Calculate the upper and lower readings on the gauage. Repeat the calculation for another type of gauge which may be treated as a second order system with natural frequency 500 Hz and damping factor 0.8.

Answers to Problems

2. $y_{ss}(t) = 1.56 - 1.02 \cos(6t - 2.3)$ approximately.

3. $y = 2e^{-2t} - 2.8e^{-t} + 0.8 \cos 2t + 0.6 \sin 2t$.

5. $\ddot{y} = 60 - 67.6 \cos(5t - 1.16)$ approximately.

6. 2310 psi and 2190 psi, approximately.
2999.9 psi and 1500.1 psi, approximately (i.e. use second guage).

3.4 THE ANALOG FILTER

Consider the electronic network shown in Figure 2.8(a) with $R = 10^5$ Ohms, $C = 2 \times 10^{-6}$ Farads and input voltage $E_1(t) = 2 \sin \pi t + 0.5 \sin 100\pi t$. Using the result of Example 2.15(a), the network is described by the transfer function

$$G(s) = \frac{1}{1 + RCs} = \frac{1}{1 + 0.2s} .$$

Now $\quad |G(jw)| = \dfrac{1}{\sqrt{1 + 0.04\, w^2}}$ and $\angle G(jw) = -\tan^{-1}(0.2w)$.

Therefore $\quad |G(j\pi)| = 0.847$, $\quad \angle G(j\pi) = -0.561$,

and $\quad |G(j100\pi)| = 0.016$, $\quad \angle G(j100\pi) = -1.555$,

and the steady-state network output is

$$E_0(t) = 1.694 \sin(\pi t - 0.561) + 0.008 \sin(100\pi t - 1.555), \text{ approx.}$$

Therefore the high frequency (50 Hz) component has been virtually removed by the action of the network. We say that the network has "filtered out" the high frequency content of the input function. There are many everyday applications where such filtering is necessary. For example, consider the measurement of pressure in a pipeline as in problem 6 of Section 3.3. Suppose that variation of pressure includes a low amplitude high frequency component in addition to the main variation of 10 Hz. The frequency response characteristics of the measuring instrument may tend to amplify this high frequency component to such an extent that it is impossible to read the instrument's display. It is therefore necessary to include an electronic network before the measuring instrument to remove the unwanted high frequency signal. A device chosen to modify the frequency characteristics of a *continuous* signal in some specified way is known as an **analog filter**. In particular the first order system $G(s) = \dfrac{1}{1 + Ts}$ passes at least 90% of input components of frequency less that $0.077T^{-1}$ Hz, and passes less that 10% of components of frequency greater than $1.584T^{-1}$ Hz. The first order system is an approximation to what is called a *low pass filter*. The gain ~ frequency diagram for an *ideal* low pass filter is shown in Figure 3.3(a). Other common filter requirements are shown in Figures 3.3(b), (c), (d). Transfer functions $G(s)$ which approximate to these ideal characteristics are well documented in the literature; an introduction to filter synthesis is given by Stanley (1975). In this section only particular examples are considered. In Chapter 7 the concept of an analog filter is used for the design of *digital filters*, i.e. filters which process *discrete* input functions.

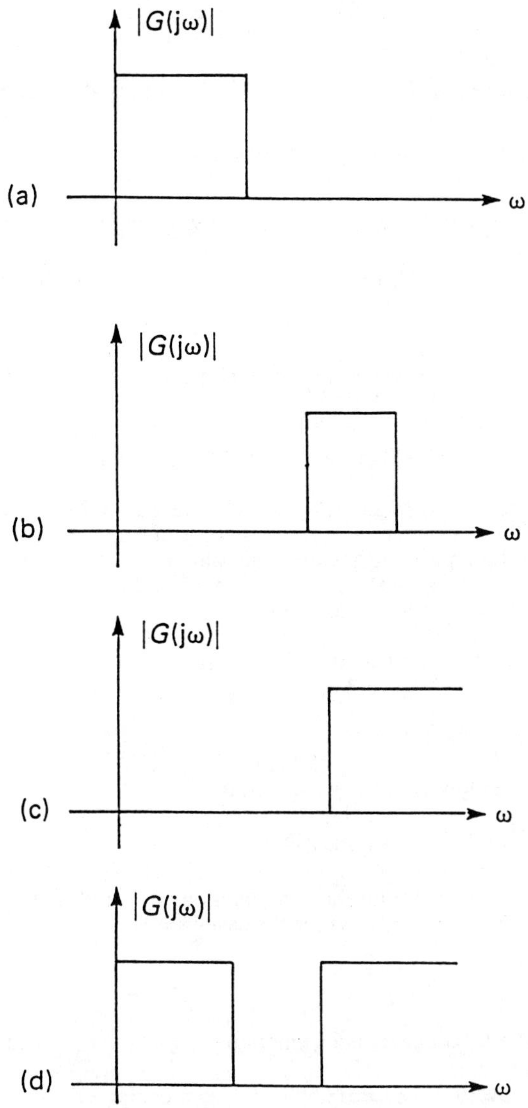

Fig. 3.3 Ideal filter gain characteristics (a) low-pass, (b) band-pass, (c) high-pass, (d) band rejection.

Example 3.6

Consider the second order system with transfer function $G(s) = \dfrac{10^5}{s^2 + 10^2 s + 10^6}$. Show that the maximum value of $|G(jw)|$ is approximately 1 and find the corresponding value of w. Obtain the frequency range for which $|G(jw)| < 0.2$. Comment on the application of this system as an analog filter.

$$G(jw) = \frac{10^5}{10^6 - w^2 + j\,10^2 w} \quad \text{and} \quad |G(jw)| = \frac{10^5}{\sqrt{(10^6 - w^2)^2 + 10^4 w^2}}$$

which will be maximised when $(10^6 - w^2)^2 + 10^4w^2$ is minimised. Differentiating gives

$$-4w(10^6 - w^2) + 2 \times 10^4 w = 0 \text{ for maximum } |G(jw)|.$$

Therefore $w^2 = 10^6 - 0.5 \times 10^4$ for which value $|G(jw| = \dfrac{10^5}{\sqrt{10^{10} - 0.25 \times 10^8}}$.

Therefore the maximum value of $|G(jw)|$ is approximately 1 and the corresponding frequency is $w = 997.5$ radians per second.

For $|G(jw)| < 0.2$ we have $\dfrac{10^{10}}{(10^6 - w^2)^2 + 10^4 w^2} < 0.04$ which can be written

$$w^4 - 1.99 \times 10^6 w^2 + 0.75 \times 10^{12} > 0$$

i.e. $w^2 < 505{,}000$ or $w^2 > 1{,}485{,}000$ approximately.

Therefore $|G(jw)| < 0.2$ for all frequencies *outside* the range $711 < w < 1219$ radians/second. The system $G(s)$ behaves as a *bandpass filter* operating in a narrow band around $w = 1000$. If the input consists of a summation of equal amplitude sine waves over a wide frequency range, then the output will be dominated by terms of frequency near $w = 1000$. The filter is said to be *tuned* to the frequency $w = 1000$.

An electronic network corresponding to the above filter can be designed using a circuit similar to the one shown in Figure 2.8(b).

Example 3.7

An approximation to the ideal low pass filter is given by

$$|G(jw|^2 = \frac{1}{1 + (w/w_c)^{2k}}$$

where k in the order of the corresponding transfer function $G(s)$, and w_c is the cut-off frequency. Obtain the transfer function $G(s)$ for the cases $k = 1, 2, 3$.

For $k = 1$, $G(s)$ is clearly the 1st order transfer function $\dfrac{1}{1 + Ts}$, $T = \dfrac{1}{w_c}$, since

$$|G(jw)| = \frac{1}{\sqrt{1 + w^2T^2}} = \frac{1}{\sqrt{1 + (w/w_c)^2}} .$$

For $k = 2$ we have $|G(jw)|^2 = \dfrac{1}{1 + (w/w_c)^4}$.

For any complex number z with conjugate $\bar{z}$ we have $z\bar{z} = |z|^2$.

Therefore $G(jw)\,G(-jw) = |G(jw)|^2 = \dfrac{1}{1 + w^4/w_c^4}$.

Recalling that $s = jw$ we have $G(s)\,G(-s) = \dfrac{w_c^4}{s^4 + w_c^4}$.

Assuming that G(s) represents a *stable* system, then all the poles of G(s) lie in the left hand half of the complex s-plane and consequently the poles of G($-s$) lie (symmetrically placed) in the right hand half-plane. Therefore we can obtain G(s) by factorising G(s) G($-s$) and selecting that part which corresponds to stable poles.

$$G(s)\,G(-s) = \frac{w_c^4}{(s+(1+j)w_c/\sqrt{2})(s+(1-j)w_c/\sqrt{2})(s+(-1+j)w_c/\sqrt{2})(s+(-1-j)w_c/\sqrt{2})} .$$

Therefore

$$G(s) = \frac{w_c^2}{(s+(1+j)w_c/\sqrt{2})(s+(1-j)w_c/\sqrt{2})} = \frac{w_c^2}{(s^2+\sqrt{2}\,w_c s+w_c^2)} .$$

Following the same method for $k = 3$,

$$G(s)\,G(-s) = \frac{w_c^6}{w_c^6 - s^6} .$$

The poles of G(s) G($-s$) are at $s = \pm w_c$,

$$s = w_c e^{\pm j\frac{\pi}{3}} = w_c\left(\frac{1}{2} \pm j\,\frac{\sqrt{3}}{2}\right) \quad \text{and}$$

$$s = w_c e^{\pm j\frac{2\pi}{3}} = w_c\left(-\frac{1}{2} \pm j\,\frac{\sqrt{3}}{2}\right).$$

Selecting only the poles in the left hand half-plane we have

$$G(s) = \frac{w_c^3}{(s+w_c)\left(s+w_c\left(\frac{1}{2}+j\,\frac{\sqrt{3}}{2}\right)\right)\left(s+w_c\left(\frac{1}{2}-j\,\frac{\sqrt{3}}{2}\right)\right)}$$

$$= \frac{w_c^3}{s^3+2w_c s^2+2w_c^2 s+w_c^3} .$$

The $|G(jw)|^2 \sim w$ graphs for $k = 1, 2, 3$ are shown in Figure 3.4.

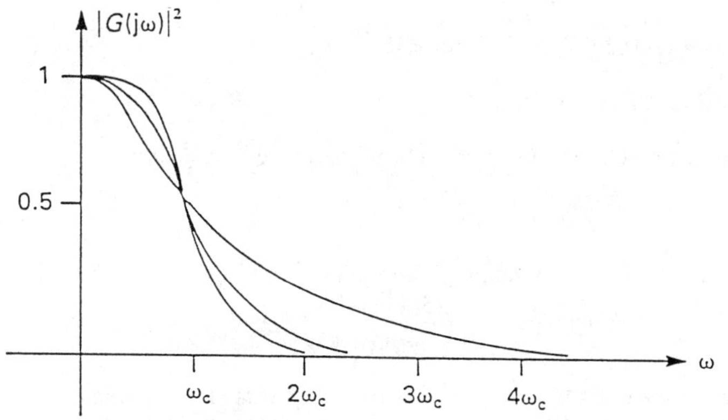

Fig. 3.4

Example 3.8

In order to stabilise a certain feedback control system it is necessary to increase the phase in the feedback path using the analog filter

$$G(s) = \frac{1 + 2Ts}{1 + Ts} \quad .$$

Calculate the largest phase advance achieved by this filter and the corresponding frequency.

$$G(jw) = \frac{1 + 2jwT}{1 + jwT} \quad \text{and} \quad \underline{|G(jw)} = \tan^{-1} 2wT - \tan^{-1} wT.$$

The maximum value for the phase is obtained from

$$\frac{d}{dw}\underline{|G(jw)} = \frac{2T}{1 + 4w^2T^2} - \frac{T}{1 + w^2T^2} = 0$$

i.e. $\quad 2(1 + w^2T^2) - (1 + 4w^2T^2) = 0.$

Therefore $w = \dfrac{1}{T\sqrt{2}}$ is the frequency corresponding to the maximum phase advance which is $\tan^{-1}\sqrt{2} - \tan^{-1}\dfrac{1}{\sqrt{2}} = 19.5°$.

Problems

1. Design a first order low pass filter G(*s*) such that

$$|G(0)| = 1,$$
$$|G(jw)| > 0.5 \quad \text{for } w < 20\pi \, ,$$
$$\text{and} \quad |G(jw)| < 0.5 \quad \text{for } w > 20\pi.$$

2. The filter $G(s) = \dfrac{s}{s^2 + s + 1}$ is a crude attempt at a band pass filter. It has the following properties.

   (i) $|G(jw)|$ is maximum when $w = w_c$,
   (ii) $|G(jw)| > 0.5\,|G(jw_c)|$ for $w_1 < w < w_2$.

   Find the values of w_c , w_1 and w_2.

3. Find the transfer function G(*s*) of the low pass filter satisfying

$$|G(jw)|^2 = \frac{1}{1 + w^8} \quad .$$

4. It is necessary to design an analog filter G(*s*) satisfying

$$|G(jw)|^2 = \frac{25(4 - w^2)^2}{(9 + w^2)(16 + w^2)} \quad .$$

   Further the filter must be stable and *minimum phase* (i.e. all poles *and zeros* of G(*s*) lie in the left hand half-plane: a minimum phase filter minimises the time delay between input and output time functions). Find G(*s*).

5. A phase-lag filter is required to modify the performance of a control system. The filter transfer function is of the form

$$G(s) = \frac{K(1 + Ts)}{1 + \lambda Ts}, \quad \lambda > 1 ,$$

and the following conditions have to be satisfied.

(i) The phase lag $-\underline{|G(jw)}$, has a maximum value of 30°.
(ii) The maximum phase lag occurs at $w = 100$ radians/second.
(iii) $|G(j\,100)| = 1$.

Find the values of λ, T and K.

Answers to problems

1. $G(s) = \dfrac{1}{1 + 0.02757s}$.

2. $w_c = 1, w_1 = 0.457 , \; w_2 = 2.189$.

3. $G(s) = \dfrac{1}{s^4 + 2.613s^3 + 3.414s^2 + 2.613s + 1}$.

4. $G(s) = \dfrac{5(s^2 + 4)}{(s+3)(s+4)}$.

5. $\lambda = 3 , T = 0.00577 , \; K = \sqrt{3}$.

3.5 RESONANCE

We have seen already the way in which a second order system can be described in terms of the parameters *natural frequency* , w_n and *damping factor* δ_n. The system may be represented by a differential equation

$$\ddot{y} + 2\delta_n w_n \dot{y} + w_n^2 y = w_n^2 x ,$$

or by a transfer function

$$G(s) = \frac{w_n^2}{s^2 + 2\delta_n w_n s + w_n^2} .$$

When the input is $x = a \sin wt$ the steady state output is, following the notation of Section 3.2,

$$y_{ss} = aR \sin(wt + \phi)$$

where $$R = |G(jw)| = \frac{w_n^2}{\sqrt{(w_n^2 - w^2)^2 + 4\delta_n^2 w_n^2 w^2}}$$

and $$\tan\phi = \tan \underline{|G(jw)} = -\frac{2\delta_n w_n w}{w_n^2 - w^2} .$$

The result of problem 7 in Section 3.2 shows that the amplification factor R is largest when $w = w_n\sqrt{1 - 2\delta_n^2}$ assuming that $\delta_n < \dfrac{1}{\sqrt{2}}$. The corresponding value of R is

$R_{MAX} = \dfrac{1}{2\delta_n\sqrt{1-\delta_n^2}} \approx \dfrac{1}{2\delta_n}$ when δ_n is small. The frequency $w_n\sqrt{1-2\delta_n^2}$ is called the **resonant frequency** of the system, i.e. the input frequency which maximises the amplitude of steady state oscillations. For a lightly damped system (δ_n small) the output amplitude can be very large indeed and may cause certain variables to reach limiting values (such as a piston in a hydraulic system hitting the end of the cylinder). In more severe cases system destruction may occur as, for example, in the Tacoma Narrows bridge disaster in 1940, see Braun (1975). In general for mechanical systems **resonance**, i.e. oscillations at resonant frequency, should be avoided by careful design. However, in certain electrical systems the phenomenon can be used to advantage (e.g. as in tuned circuits).

The analysis of lightly damped systems is often carried out by assuming them to have *zero* damping. This assumption greatly simplifies the analysis for high order systems, see Huntley and Johnson (1983). The undamped second order system has transfer function $G(s) = \dfrac{w_n^2}{s^2 + w_n^2}$, the resonant frequency is $w = w_n$ and it follows that $|G(jw_n)|$ is *infinite*. For input $x = a \sin w_n t$ the output is

$$y = L^{-1}\left\{\frac{a w_n^3}{(s^2 + w_n^2)^2}\right\} = \frac{a}{2}(\sin w_n t - w_n t \cos w_n t)$$

and it is clear that the output amplitude becomes infinite as $t \to \infty$. Higher order systems which are *marginally stable* (i.e. $G(s)$ has a *simple* pole on the imaginary axis) also exhibit this property. The denominator of $G(s)$ will have a non-repeated factor of the form $s^2 + k^2$ and for input $x = a \sin kt$ the amplitude of output oscillations will tend to infinity. For lightly damped systems the natural frequency and resonant frequency are virtually the same for each mode of oscillation. The assumption of zero damping allows the *critical frequencies* to be more readily obtained.

Example 3.9

Write down the approximate steady state output $y_{ss}(t)$ for the system described by the differential equation

$$\ddot{y} + 16\dot{y} + 40000y = 120000 \sin 200t .$$

If we take the input to be $x = 3 \sin 200t$, then the system transfer function is

$$G(s) = \frac{40000}{s^2 + 16s + 40000} .$$

Therefore $w_n = 200$ and $\delta_n = 0.04$. Since δ_n is very small the natural frequency and resonant frequency are virtually equal, and the amplification factor is given by $\dfrac{1}{2\delta_n} = 12.5$. The phase at the natural frequency is –90° (see Figure 3.2) and therefore the steady state output is

$$y_{ss} = 12.5 \times 3 \sin\left(200t - \frac{\pi}{2}\right) = -37.5 \cos 200t .$$

[Alternatively we can write $G(j\,200) = \dfrac{40000}{3200j} = \dfrac{12.5}{j}$

Therefore $R = 12.5$ and $\phi = -90°$.]

Example 3.10

Consider the double spring system shown in Figure 3.5. Assuming linear springs and very small damping, obtain an approximation to the transfer function $\dfrac{L\{y_1\}}{L(x)\}}$. If the point A is moved up and down at frequency w radians/second, find approximate values for the critical frequencies for the case

$$\frac{k_1}{m_1} = 12\ , \ \frac{k_2}{m_1} = \frac{k_2}{m_2} = 8.$$

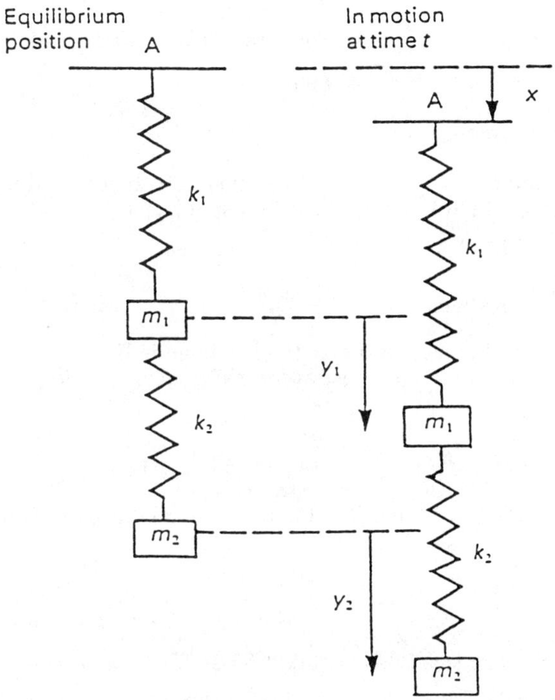

Fig. 3.5 k_1, k_2 are the spring stiffnesses; m_1, m_2 are masses.

Making the simplifying assumption of zero damping and applying Newton's second law to each mass gives

$$m_1\ddot{y}_1 = k_2(y_2 - y_1) - k_1(y_1 - x)$$

$$m_2\ddot{y}_2 = -k_2(y_2 - y_1)\ .$$

Taking Laplace transforms (assuming zero initial conditions)

$$(m_1s^2 + k_1 + k_2)\,L\{y_1\} - k_2\,L\{y_2\} = k_1\,L\{x\}$$
$$-\,k_2\,L\{y_1\} + (m_2s^2 + k_2)\,L\{y_2\} = 0.$$

Therefore

$$((m_1s^2 + k_1 + k_2)(m_2s^2 + k_2) - k_2^2)\,L\{y_1\} = k_1\,(m_2s^2 + k_2)\,L\{x\}$$

and

$$\frac{L\{y_1\}}{L\{x\}} = \frac{k_1/m_1\,(s^2 + k_2/m_2)}{s^4 + (k_1/m_1 + k_2/m_1 + k_2/m_2)\,s^2 + (k_1/m_1)(k_2/m_2)}\,.$$

Using the given numbers the denominator of the transfer function is

$$s^4 + 28s^2 + 96 = (s^2 + 4)(s^2 + 24).$$

Therefore the critical frequencies are $w = 2$ and $w = \sqrt{24}$. Since the damping is small, the critical frequencies for the damped system will be very near these values.

Problems

1. Write down the approximate steady state solution of the differential equation

$$\ddot{y} + 0.5\dot{y} + 100y = 7 \sin nt$$

where n is the resonant frequency.

2. The position of a component in a mechanical system subject to vibration at frequency w is given by the solution of the differential equation

$$\dddot{x} + 9140\ddot{x} + 1.3 \times 10^7 x = 1000 \sin wt\,.$$

What are the critical frequencies for the system?

3. For the circuit shown in Figure 3.6 the applied voltage is $E(t) = 250 \sin 300t$ volts. Show that the current i amps satisfies the differential equation

$$L\frac{d^2i}{dt^2} + R\frac{di}{dt} + \frac{i}{C} = \frac{dE}{dt}\,.$$

Given that $\sqrt{\frac{1}{LC}} = 300$, $L = 11$ and $R = 40$, will a 13 amp fuse be adequate for this circuit?

4. Show that the system $G(s) = \dfrac{100}{s^2 + 2s + 100}$ with input $x = 5 \sin wt$ at resonant frequency, has approximate steady-state output $y = -25 \cos wt$. If the input is filtered using a first order filter $\dfrac{1}{1 + sT}$, find the value of T which reduces the amplification factor to unity.

5. A simplified model of a car travelling along a bumpy road is shown in Figure 3.7. Suppose that the mass M of the car and spring stiffness k are such that $M/k = 0.01s^2$. Neglecting damping show that the displacement $y(t)$ of the car from its equilibrium position satisfies the differential equation

$$0.01\ddot{y} + y = h \cos\left(\frac{2\pi Vt}{\lambda}\right)$$

where V is the constant speed of the car and the road is assumed to have cosine wave profile of amplitude h and wavelength λ. Deduce that for a road of wavelength 15m, the worst vibrations occur when $V \approx 24$ m/s.

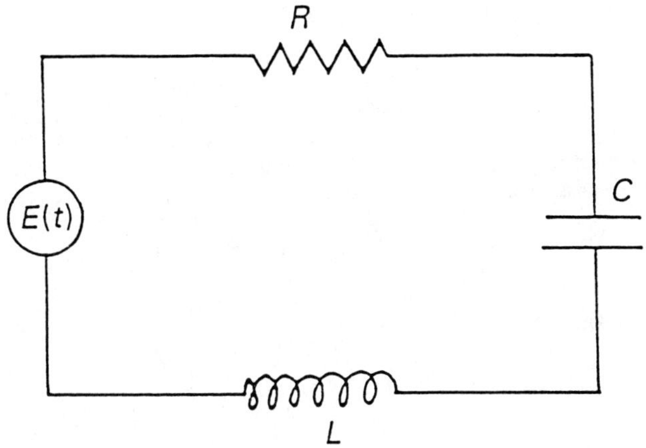

Fig. 3.6

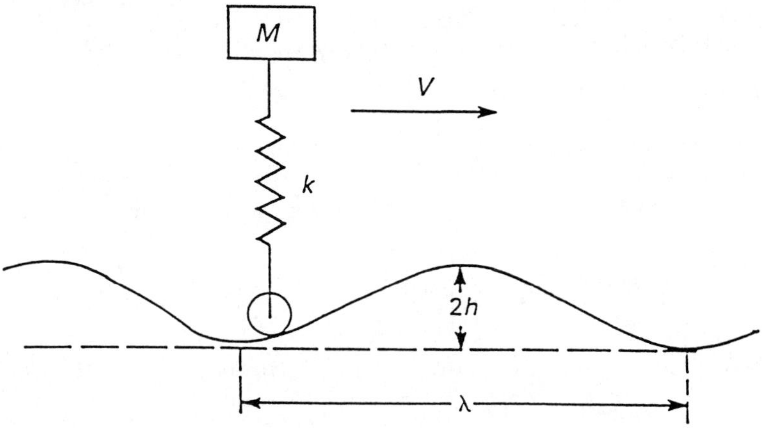

Fig. 3.7

Answers to problems

1. $y_{ss} \approx -1.4 \cos 10t$.

2. 42, 85.9 radians/second.

3. Yes, steady state current has amplitude $\approx$ 6.25 amps.

4. $T \approx 0.49$.

4

Piecewise Continuous Functions

4.1 INTRODUCTION

In earlier chapters input functions, or alternatively functions on the right-hand side of differential equations, have mostly been restricted to those which can be described by a simple continuous function for $t \geq 0$, e.g. $x = t$, $x = \sin wt$, etc. In this chapter we consider inputs which require a multiple definition and may have discontinuities. For example the **unit pulse function** $p_\alpha(t)$ is defined as

$$p_\alpha(t) = \begin{cases} 1 & \text{when } 0 \leq t < \alpha \\ 0 & \text{when } t \geq \alpha \end{cases} ,$$

and the function defined by

$$x(t) = \begin{cases} t & \text{when } 0 \leq t < 1 \\ 1 & \text{when } t \geq 1 \end{cases}$$

is a *limited* form of the ramp function.

Laplace transform techniques will be applied to this type of input and considerable use will be made of the **shift theorem.** The concepts of pulse functions, delayed functions and impulse functions mentioned briefly in Chapter 1 will be studied here in more detail. One example deals with a sequence of narrow width pulses and serves as an introduction to sampling devices which are considered in Chapter 5.

4.2 THE DELAY OPERATION

We define the *delayed* unit step function, $u(t - \mathrm{T})$, as follows:-

$$u(t - \mathrm{T}) = \begin{cases} 1 & \text{when } t \geq \mathrm{T} \\ 0 & \text{when } t < \mathrm{T} \end{cases} .$$

Delayed versions of other functions defined for $t \geq 0$ can be expressed in terms of $u(t - \mathrm{T})$.

The function $f(t - \mathrm{T})u(t - \mathrm{T})$ represents $f(t)$ delayed by amount T, i.e. shifted T units to the right along the horizontal axis. Figure 4.1 shows the **delayed ramp function** $(t - \mathrm{T})u(t - \mathrm{T})$, which is the result of shifting $tu(t)$ T units in the positive t direction.

The unit impulse function $\delta(t)$ and the delayed unit impulse function $\delta(t - \mathrm{T})$ have already been defined in Chapter 1. Here it is convenient to redefine them in terms of the *unit pulse function* $p_\alpha(t)$. Thus

$$p_\alpha(t) = \begin{cases} 1 & \text{when } 0 \leq t < \alpha \\ 0 & \text{otherwise} \end{cases} ,$$

$$\delta(t) = \lim_{\alpha \to 0} \frac{p_\alpha(t)}{\alpha} ,$$

$$\delta(t - \mathrm{T}) = \lim_{\alpha \to 0} \frac{p_\alpha(t - \mathrm{T})}{\alpha} .$$

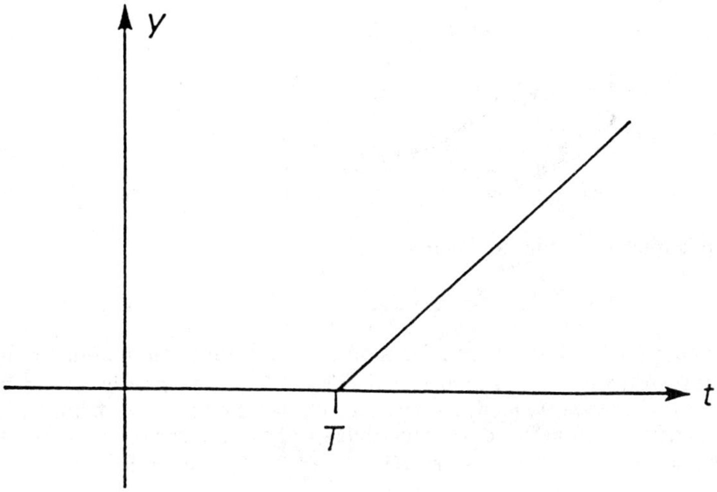

Fig. 4.1 The function $y = (t - T)u(t - T)$.

We also define the **delayed unit pulse function** $p_\alpha(t - \mathrm{T})$ as

$$p_\alpha(t - \mathrm{T}) = \begin{cases} 1 & \text{when } \mathrm{T} \le t < \mathrm{T} + \alpha \\ 0 & \text{otherwise} \end{cases} .$$

Note that $p_\alpha(t)$ and $p_\alpha(t - \mathrm{T})$ can be conveniently defined in terms of step functions as follows:-

$$\begin{aligned} p_\alpha(t) &= u(t) - u(t - \alpha) \\ p_\alpha(t - \mathrm{T}) &= u(t - \mathrm{T}) - u(t - \mathrm{T} - \alpha). \end{aligned}$$

From the above definitions and the definition of the Laplace transform it is easy to verify that

$$\begin{aligned} \mathrm{L}\{u(t - \mathrm{T})\} &= \frac{e^{-sT}}{s} , \\ \mathrm{L}\{p_\alpha(t)\} &= \frac{1}{s}\,(1 - e^{-s\alpha}) , \\ \mathrm{L}\{p_\alpha(t - \mathrm{T})\} &= \frac{e^{-sT}}{s}\,(1 - e^{-s\alpha}) , \\ \mathrm{L}\{\delta(t)\} &= 1 , \\ \mathrm{L}\{\delta(t - \mathrm{T})\} &= e^{-sT} . \end{aligned}$$

Some of the above results are special cases of the following theorem.

Theorem 4.1 (Shift Theorem)

$L\{f(t-T)u(t-T)\} = e^{-sT}L\{f(t)\}$

Proof

$$\begin{aligned} L\{f(t-T)u(t-T)\} &= \int_T^{\infty} f(t-T)e^{-st}\,dt \\ &= \int_0^{\infty} f(\tau)e^{-s(\tau+T)}\,d\tau\ , \quad (\tau = t-T) \\ &= e^{-sT}\int_0^{\infty} f(\tau)e^{-s\tau}\,d\tau \\ &= e^{-sT}L\{f(t)\}\ . \end{aligned}$$

The shift theorem is most useful in its *inverse* form,

$$L^{-1}\{e^{-sT}F(s)\} = f(t-T)u(t-T)\ , \quad \text{where } F(s) = L\{f(t)\}\ .$$

It follows from Theorem 4.1 that the transform of a delayed function is obtained by multiplying the transform of the original function by e^{-sT}. The term e^{-sT} can be thought of as the *transfer function* of a system which has the effect of delaying its input by a time T, and is often referred to as the **delay operation**. For example, if the ramp function $f(t) = t$ is input to the system $G(s) = e^{-sT}$, then the output would be $L^{-1}\left\{\frac{e^{-sT}}{s^2}\right\} = (t-T)u(t-T)$, the *delayed* ramp function.

Example 1.1

The system with transfer function $G_1(s) = e^{-sT}$ has input $x(t) = a\sin wt$. Write down the system output. What is the output of the system $G(s) = \frac{e^{-sT}}{1+s}$ when the input is $a\sin wt$.

The Laplace transform of the output of $G_1(s)$ is $\frac{awe^{-sT}}{s^2+w^2}$.

It follows from the inverse form of the shift theorem that the system output is

$$a\sin w(t-T)u(t-T).$$

Note that for $t<T$ the output is zero and for $t>T$ the output is $a\sin w(t-T)$. This "steady state" output can be obtained from the frequency response function e^{-jwT} for the system $G_1(s)$. $|G_1(jw)| = 1$ and $\underline{|G(jw)} = -wT$, and therefore the steady state output is $a\sin(wt-wT)$.

For the system $G(s) = \frac{e^{-sT}}{1+s}$ the output $y(t)$ is obtained from

$$\begin{aligned} L\{y(t)\} &= e^{-sT}\left(\frac{aw}{(s+1)(s^2+w^2)}\right) \\ &= \frac{ae^{-sT}}{1+w^2}\left(\frac{w}{s+1} - \frac{ws-w}{s^2+w^2}\right). \end{aligned}$$

Therefore

$$y(t) = \frac{a}{1+w^2}\left(we^{-(t-T)} - w\cos w(t-T) + \sin w(t-T)\right)u(t-T)\,.$$

That is

$$y(t) = \begin{cases} 0 & \text{when } t < T \\ \dfrac{a}{1+w^2}\left(we^{-(t-T)} - w\cos w(t-T) + \sin w(t-T)\right) & \text{when } t \geq T \end{cases}\,.$$

Example 4.2

Express the function shown in Figure 4.2 in terms of the ramp function and a delayed function, and hence write down its Laplace transform.

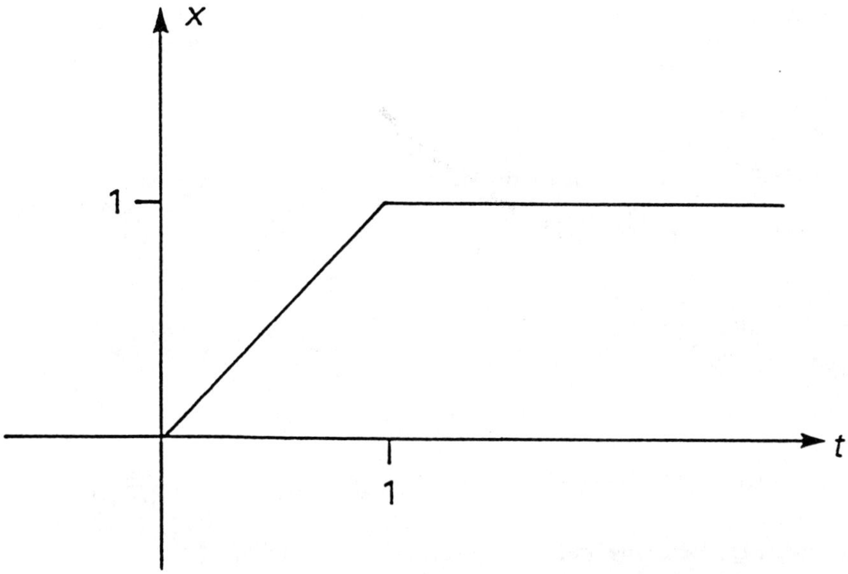

Fig. 4.2

For $0 \leq t < 1$, the function is $x(t) = t$.

For $t \geq 1$, $x(t) = 1 = t - (t-1)$.

Therefore $x(t) = tu(t) - (t-1)\,u(t-1)$. This function is known as the **limited unit ramp function**.

$$L\{x(t)\} = \frac{1}{s^2} - \frac{1}{s^2}\, e^{-s} \ .$$

This result can be confirmed directly using the definition of the Laplace transform.

Example 4.3

Obtain the Laplace transform of the function

$$x(t) = \begin{cases} 2-t & 0 \le t < 1 \\ 0 & \text{otherwise} \end{cases} .$$

$x(t)$ may be written in the form $x(t) = (2-t)\,u(t) - (2-t)\,u(t-1)$ but the 2nd term is not in the special form $f(t-\mathrm{T})u(t-\mathrm{T})$ required to apply Theorem 4.1.

Therefore we write

$$\begin{aligned} x(t) &= (2-t)u(t) - (1 + (1-t))u(t-1) \\ &= (2-t)u(t) + (t-1)u(t-1) - u(t-1) \end{aligned}$$

and applying the shift theorem

$$L\{x(t)\} = \frac{2}{s} - \frac{1}{s^2} + \frac{e^{-s}}{s^2} - \frac{e^{-s}}{s} \ .$$

Problems

1. (a) Determine the Laplace transform of the following functions.

   (i) $f(t) = (t-2)^2u(t-2)$ (ii) $f(t) = t\,u(t-1)$

   (iii) $f(t) = \begin{cases} 0 & , \ t < 1 \\ 2 & , \ 1 \le t < 2 \\ 1 & , \ t \ge 2 \end{cases} .$

   (b) Determine the following inverse Laplace transforms

   (i) $L^{-1}\left\{\dfrac{e^{-s}}{(s+1)^2}\right\}$ (ii) $L^{-1}\left\{\dfrac{e^{-2s}}{s(s+1)}\right\}$.

2. Find the Laplace transform of the functions

   (i) $\cos\left(t - \dfrac{2\pi}{3}\right)u\left(t - \dfrac{2\pi}{3}\right)$.

   (ii) $f(t) = \begin{cases} 0 & \text{when } t < a \\ t-a & \text{when } a \le t < b \\ b-a & \text{when } t \ge b \end{cases} .$

3. For the function $f(t)$ defined in Problem 2(ii) write down the transform of $f\,'(t)$ and show that

$$\lim_{b \to a}\left(\frac{L\{f\,'(t)\}}{(b-a)}\right) = e^{-as} \ . \text{ Explain.}$$

4. Obtain the inverse Laplace transform of

(i) $(1 - e^{-sT})\left(\frac{1}{s} - \frac{1}{(s+1)^2}\right)$ (ii) $e^{-s}\left(\frac{e^{-2s}}{s^2} - \frac{e^{-4s}}{s^2}\right)$.

5. The function $x(t)$ is defined as follows:-

$$x(t) = \begin{cases} t & \text{for } 0 \le t < 1 \\ 0 & \text{for } t \ge 1 \end{cases}.$$

Using delayed step functions to express $x(t)$, write down its Laplace transform. Find the output of the system $G(s) = \frac{1}{1 + 0.1s}$ when the input is $x(t)$.

6. Find the Laplace transform of the staircase function

$$f(t) = n + 1 \quad \text{when } nT \le t < (n+1)T, \quad n = 0, 1, 2, \ldots .$$

If $f(t)$ is input to the system $G(s) = \frac{1}{1+s}$ obtain the output in the interval $nT \le t < (n+1)T$.

Answers to Problems

1. (a) (i) $\frac{2e^{-2s}}{s^3}$ (ii) $\frac{e^{-s}}{s^2} + \frac{e^{-s}}{s}$ (iii) $\frac{2e^{-s}}{s} - \frac{e^{-2s}}{s}$

(b) (i) $(t-1)e^{-(t-1)}u(t-1)$ (ii) $(1 - e^{-(t-2)})u(t-2)$.

2. (i) $\frac{e^{-2\pi s/3}s}{s^2+1}$ (ii) $\frac{1}{s^2}(e^{-as} - e^{-bs})$.

3. $\frac{1}{s}(e^{-as} - e^{-bs})$. $\frac{f'(t)}{b-a}$ is a unit area pulse function which tends to $\delta(t-a)$ as $b \to a$.

4. (i) $f(t) = \begin{cases} 1 - te^{-t} & \text{when } 0 \le t < T \\ (t-T)e^{-(t-T)} - te^{-t} & \text{when } t \ge T \end{cases}$.

(ii) $(t-3)u(t-3) - (t-5)u(t-5) = \begin{cases} 0 & \text{when } 0 \le t < 3 \\ t-3 & \text{when } 3 \le t < 5 \\ 2 & \text{when } t \ge 5 \end{cases}$.

5. $x(t) = t - (t-1)u(t-1) - u(t-1)$,

$$L\{x(t)\} = \frac{1}{s^2}(1 - e^{-s}) - \frac{e^{-s}}{s},$$

$$\left(t - \frac{1}{10} + \frac{1}{10}e^{-10t}\right)u(t) - \left(t - \frac{1}{10} - \frac{9}{10}e^{-10(t-1)}\right)u(t-1).$$

6. $\frac{1}{s}(1 + e^{-sT} + e^{-s2T} + e^{-s3T} + \ldots\ldots\ldots),$

$\sum_{i=0}^{n} (1 - e^{-(t - iT)})$.

4.3 PERIODIC FUNCTIONS

The periodic function $f_p(t)$ defined for $t \geq 0$ by

$$f_p(t) = f(t) \text{ for } 0 \leq t < \mathrm{L} \ , \ f_p(t + \mathrm{L}) = f_p(t) \ ,$$

can easily be expressed in terms of delayed step functions as follows:

$$f_p(t) = \sum_{n=0}^{\infty} f(t - n\mathrm{L})(u(t - n\mathrm{L}) - u(t - [n+1]\mathrm{L})) \ .$$

Hence the Laplace transform of $f_p(t)$ is

$$\begin{aligned} \mathrm{L}\{f_p(t)\} &= \sum_{n=0}^{\infty} \int_{nL}^{(n+1)L} f(t - n\mathrm{L})e^{-st}\, dt \\ &= \sum_{n=0}^{\infty} \int_0^L f(\tau)\, e^{-s(\tau + nL)}\, d\tau \\ &= \left(\int_0^L f(\tau)\, e^{-s\tau} d\tau\right) \sum_{n=0}^{\infty} e^{-snL} \qquad (4.1) \end{aligned}$$

Theorem 4.2

The Laplace transform of the periodic function $f_p(t)$ of period L is given by

$$\mathrm{L}\{f_p(t)\} = \frac{1}{1 - e^{-sL}} \int_0^L f_p(t)\, e^{-st}\, dt \ .$$

Proof

The proof follows from equation (4.1) by summing the geometric series (assuming that the real part of s is positive) and noting that $f_p(t) = f(t)$ for $0 \leq t < \mathrm{L}$.

Example 4.4

Find the Laplace transform of the half-wave rectified signal

$$f_p(t) = \begin{cases} \sin t & \text{when } 0 \leq t < \pi \\ 0 & \text{when } \pi \leq t < 2\pi \end{cases} \ , \ f_p(t + 2\pi) = f_p(t).$$

From Theorem 4.2

$$\begin{aligned} \mathrm{L}\{f_p(t)\} &= \frac{1}{1 - e^{-s2\pi}} \int_0^{2\pi} f_p(t) e^{-st}\, dt \\ &= \frac{1}{1 - e^{-s2\pi}} \int_0^{\pi} \sin t\, e^{-st}\, dt \end{aligned}$$

$$= \frac{1}{1-e^{-s2\pi}}\left(\frac{1+e^{-s\pi}}{s^2+1}\right)$$
$$= \frac{1}{(1-e^{-s\pi})(s^2+1)} .$$

Note that this Laplace transform can be expressed as an infinite series

$$L\{f_p(t)\} = (1 + e^{-s\pi} + e^{-s2\pi} + e^{-s3\pi} + \ldots\ldots)\frac{1}{1+s^2} ,$$

and using Theorem 4.1 we have

$$f_p(t) = \sin t + \sin(t-\pi)u(t-\pi) + \sin(t-2\pi)\,u(t-2\pi) + \ldots\ldots$$
$$= \sin t\,(1 - u(t-\pi) + u(t-2\pi) - u(t-3\pi) + \ldots\ldots)$$

as an alternative representation of the given function.

Example 4.5

Find the Laplace transform of the square wave function

$$x_p(t) = \begin{cases} 1 & \text{when } 0 \le t < L \\ 0 & \text{when } L \le t < 2L \end{cases} , \quad x_p(t+2L) = x_p(t) .$$

If $x_p(t)$ is the input to a system with transfer function $G(s) = \frac{1}{1+sT}$ obtain the system output $y(t)$.

$$L\{x_p(t)\} = \frac{1}{1-e^{-s2L}}\int_0^L e^{-st}\,dt$$
$$= \frac{1}{1-e^{-s2L}}\left(\frac{1-e^{-sL}}{s}\right)$$
$$= \frac{1}{(1+e^{-sL})s}$$
$$= (1 - e^{-sL} + e^{-s2L} - e^{-s3L} + \ldots\ldots)\frac{1}{s} .$$

[This result can be obtained directly by simply expressing $x_p(t)$ in terms of delayed step functions.]

The output from the system $G(s)$ is given by

$$L\{y(t)\} = (1 - e^{-sL} + e^{-s2L} - e^{-s3L} + \ldots\ldots)\frac{1}{s(1+sT)}$$
$$= (1 - e^{-sL} + e^{-s2L} - e^{-s3L} + \ldots\ldots)\left(\frac{1}{s} - \frac{T}{1+sT}\right)$$

Therefore

$$y(t) = (1 - e^{-t/T})u(t) + \sum_{r=1}^{\infty}(-1)^r\,(1 - e^{-(t-rL)/T})\,u(t-rL).$$

Problems

1. Find the Laplace transform of the sawtooth waveform defined by the function $f(t)$,

$$f(t) = \frac{kt}{T}\,,\quad 0 \le t < T,\ f(t+T) = f(t).$$

2. Show that the Laplace transform of the full rectified wave, $f(t) = |\cos t|$, is given by

$$F(s) = \frac{s + \left(\sinh \frac{\pi s}{2}\right)^{-1}}{s^2+1}\,.$$

3. Obtain the Laplace transform of the function

$$f(t) = \begin{cases} 1 & \text{when } 0 \le t < \frac{L}{2} \\ -1 & \text{when } \frac{L}{2} \le t < L \end{cases}\,,\quad f(t+L) = f(t)\,.$$

Hence write down the Laplace transform of the triangular waveform

$$v(t) = \begin{cases} t & \text{when } 0 \le t < \frac{L}{2} \\ L - t & \text{when } \frac{L}{2} \le t < L \end{cases}\,,\quad f(t+T) = f(t)\,.$$

Answers to Problems

1. $\dfrac{k}{Ts^2} - \dfrac{ke^{-sT}}{s(1-e^{-sT})}\,.$

3. $F(s) = \dfrac{1}{s}\tanh\dfrac{Ls}{4}\,,\quad V(s) = \dfrac{1}{s^2}\tanh\dfrac{Ls}{4}\,.$

4.4 SOLUTIONS OF DIFFERENTIAL EQUATIONS WITH PIECEWISE CONTINUOUS INPUTS

The output of a system for a given piecewise continuous input has already been considered briefly in Section 4.2 (problems 5 and 6) and Section 4.3. In this section a number of examples are presented where the system (or differential equation) input is a piecewise continuous function. The input is expressed in terms of delayed step functions or impulses so that its Laplace transform is readily obtained.

Example 4.6

Solve the initial value problem

$$\ddot{y} + 4y = 4u(t-2)\,,\quad y(0) = 10,\ \dot{y}(0) = 0.$$

Taking Laplace transforms gives

$$s^2 L\{y\} - s10 + 4L\{y\} = \frac{4e^{-2s}}{s}\,.$$

Therefore
$$L\{y\} = \frac{10s}{s^2+4} + \frac{4e^{-2s}}{s(s^2+4)}$$
$$= \frac{10s}{s^2+4} + \left(\frac{1}{s} - \frac{s}{s^2+4}\right) e^{-2s} .$$

Therefore, using Theorem 4.1
$$y = 10 \cos 2t + (1 - \cos 2(t-2))\, u(t-2)$$

or alternatively
$$y = \begin{cases} 10 \cos 2t & \text{when } t < 2 \\ 10 \cos 2t + 1 - \cos 2(t-2) & \text{when } t \geq 2 \end{cases} .$$

Example 4.7

The motion of a simple mechanical system is modelled by the differential equation
$$\ddot{y} + 2\dot{y} + y = x(t) , \quad y(0) = \dot{y}(0) = 0 ,$$

where the input function $x(t)$ is the limited ramp function defined by
$$x(t) = \begin{cases} t & \text{when } 0 \leq t < 1 \\ 1 & \text{when } t \geq 1 \end{cases} .$$

Obtain the function $y(t)$ for $t \geq 1$.

$$x(t) = t\, u(t) - (t-1)\, u(t-1), \text{ and}$$
$$L\{x(t)\} = \frac{1}{s^2}(1 - e^{-s}) .$$

Transforming the differential equation gives
$$L\{y\} = \frac{1}{s^2(s+1)^2}(1 - e^{-s})$$
$$= \left(\frac{1}{s^2} - \frac{2}{s} + \frac{1}{(s+1)^2} + \frac{2}{(s+1)}\right)(1 - e^{-s}) .$$

Therefore
$$y(t) = (t - 2 + te^{-t} + 2e^{-t})u(t) - (t - 3 + (t-1)e^{-(t-1)} + 2e^{-(t-1)})u(t-1)$$

and for $t \geq 1$
$$y(t) = 1 + (t+2)e^{-t} - (t+1)e^{-(t-1)} .$$

Example 4.8

Solve $\dot{y} + y = \text{K}p_\alpha(t)$, $y(0) = 0$, where $p_\alpha(t)$ is the unit pulse function and K is a constant. Discuss the form of the solution when α is small, and the result of approximating $\text{K}p_\alpha(t)$ by a suitable impulse function.

Transforming $\dot{y} + y = \text{K}p_\alpha(t)$, $y(0) = 0$,

we obtain
$$sL\{y\} + L\{y\} = \text{K}\left(\frac{1 - e^{-s\alpha}}{s}\right), \text{ see Section 4.2.}$$

Therefore
$$L\{y\} = \frac{K}{s(s+1)} - \frac{Ke^{-s\alpha}}{s(s+1)}$$
$$= K\left(\frac{1}{s} - \frac{1}{(s+1)}\right) - Ke^{-s\alpha}\left(\frac{1}{s} - \frac{1}{(s+1)}\right).$$

Therefore
$$y = K(1 - e^{-t})u(t) - K(1 - e^{-(t-\alpha)})\,u(t-\alpha)$$

i.e.
$$y = \begin{cases} K(1 - e^{-t}) & \text{for } 0 \le t < \alpha \\ K(e^{-(t-\alpha)} - e^{-t}) & \text{for } t \ge \alpha \end{cases}.$$

Figure 4.3 shows the graph of the solution with $K(1 - e^{-\alpha})$ as the peak value of y.

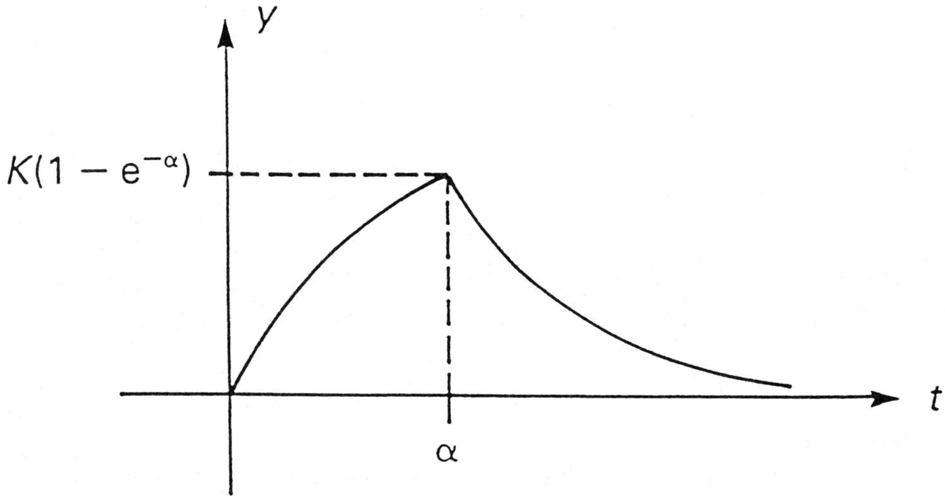

Fig. 4.3

If α is small then the peak value is approximately $K\alpha$

$$\left(\text{since } 1 - e^{-\alpha} = 1 - 1 + \alpha - \frac{\alpha^2}{2} + \frac{\alpha^3}{6} - \ldots\ldots\ldots\right),$$

and $K\alpha$ is the "area" of the pulse function $Kp_{\alpha}(t)$. If the pulse function had been $2Kp_{0.5\alpha}(t)$ the area is still $K\alpha$ and the peak value of y for small α is still approximately $K\alpha$, see Figures 4.4(a) and 4.4(b).

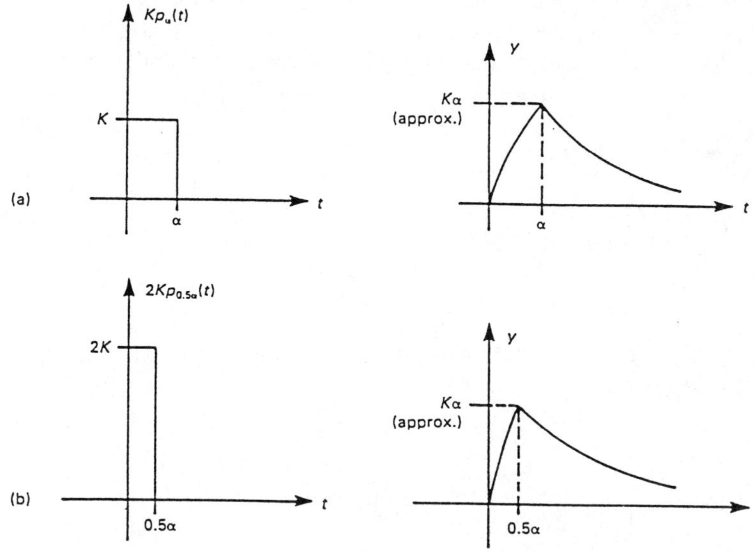

Fig. 4.4

When α is small it is clear that the important parameter is the *area* of the pulse and not its height. Thus for small α we could approximate to $Kp_{\alpha}(t)$ using $K\alpha\delta(t)$ since the area of the impulse $\delta(t)$ is 1 by definition.

Solving $\dot{y} + y = K\alpha\delta(t)$, $y(0) = 0$, we obtain after taking Laplace transforms

$$L\{y\} = \frac{K\alpha}{s+1}.$$

Therefore $y = K\alpha e^{-t}$ is the approximate solution (see Figure 4.5).

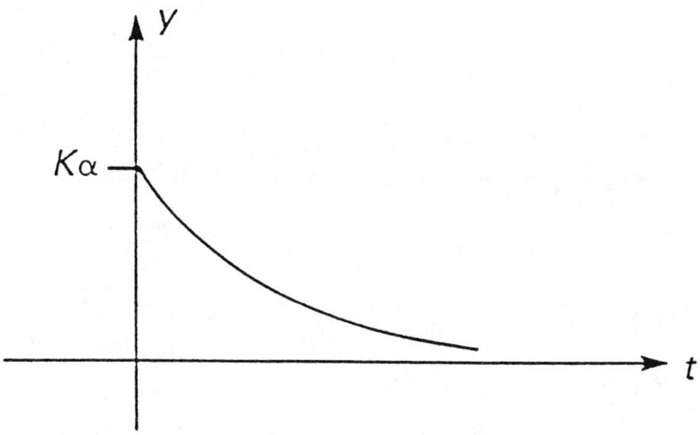

Fig. 4.5

Comparison of Figures 4.4(a) and 4.5 shows that the $\alpha\delta(t)$ approximation for $p_\alpha(t)$ is good for small α. The advantage of using $\delta(t)$ is that its Laplace transform is 1 which reduces the amount of algebraic manipulation required to complete the solution.

Example 4.9

Solve $\quad \dot{y} + y = f(t)\ ,\ y(0) = 0$

where $\quad f(t) = \begin{cases} 10^4 & \text{when } 0 \le t < 0.001 \\ 0 & \text{when } t \ge 0.001 \end{cases}$.

Since the duration of the pulse is small we can use the $\delta(t)$ approximation. The "area" of the pulse is $10^4 \times 0.001 = 10$, and we solve

$$\dot{y} + y = 10\delta(t)\ ,\ y(0) = 0.$$

Therefore $\quad L\{y\} = \dfrac{10}{s+1}$ and $y = 10e^{-t}u(t)$, approximately.

[The reader is invited to solve the given equation without approximation and to carefully compare the two solutions by plotting accurate graphs.]

Example 4.10

Consider the electronic system shown in Figure 4.6.

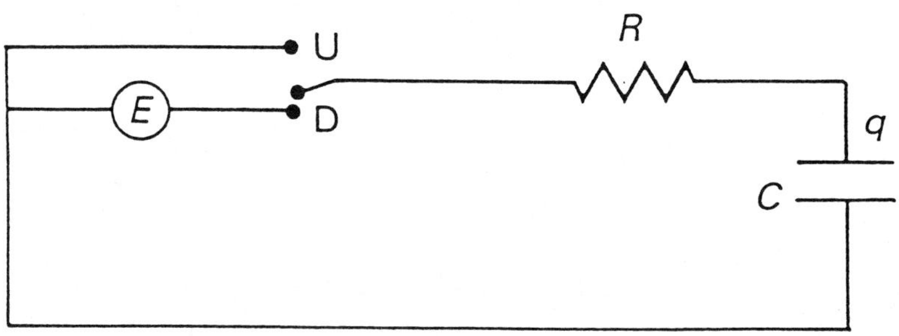

Fig. 4.6

The differential equation modelling this system is

$$\frac{dq}{dt} + \frac{q}{RC} = \begin{cases} 0 & \text{when switch at U} \\ \dfrac{E}{R} & \text{when switch at D} \end{cases}$$

where q is the charge on the capacitor, and E is a constant voltage. Assuming that $q(0) = 0$ solve the differential equation for each of the following situations.

(a) Switch set to D at $t = 0$,
(b) switch set to D at $t = 0$ and returned to U at $t = 2$,
(c) switch set to D at $t = 0$ and returned to U at $t = \alpha$ where α is small,
(d) switch set to D for short duration α at the times $t = 0$, T, 2T, 3T, ……

We write the differential equations as

$$\dot{q} + \frac{q}{\mathrm{RC}} = f(t) \quad \text{and interpret } f(t)$$

for the four cases.

(a) $f(t) = \frac{\mathrm{E}}{\mathrm{R}} u(t)$ and $\mathrm{L}\{f(t)\} = \frac{\mathrm{E}}{\mathrm{R}s}$.

Therefore $\left(s + \frac{1}{\mathrm{RC}}\right)\mathrm{L}\{q\} = \frac{\mathrm{E}}{\mathrm{R}s}$

i.e. $\mathrm{L}\{q\} = \frac{\mathrm{E}}{\mathrm{R}}\left(\frac{1}{s\left(s + \frac{1}{\mathrm{RC}}\right)}\right) = \frac{\mathrm{E}}{\mathrm{R}}\left(\frac{\mathrm{RC}}{s} - \frac{\mathrm{RC}}{\left(s + \frac{1}{\mathrm{RC}}\right)}\right)$.

Therefore $q = \mathrm{EC}(1 - e^{-t/RC})u(t)$.

(b) $f(t) = \frac{\mathrm{E}}{\mathrm{R}}(u(t) - u(t-2))$ and $\mathrm{L}\{f(t)\} = \frac{\mathrm{E}}{\mathrm{R}s}(1 - e^{-2s})$.

Therefore $\mathrm{L}\{q\} = \frac{\mathrm{E}}{\mathrm{R}}\left(\frac{1}{s\left(s + \frac{1}{\mathrm{RC}}\right)} - \frac{e^{-2s}}{s\left(s + \frac{1}{\mathrm{RC}}\right)}\right)$

and $q = \mathrm{EC}(1 - e^{-t/RC})u(t) - \mathrm{EC}(1 - e^{-(t-2)/RC})\, u(t-2)$

i.e. $q = \begin{cases} \mathrm{EC}(1 - e^{-t/RC}) & \text{when } 0 \le t < 2 \\ \mathrm{EC}(e^{-(t-2)/RC} - e^{-t/RC}) & \text{when } t \ge 2 \end{cases}$.

(c) $f(t) = \frac{\mathrm{E}\alpha}{\mathrm{R}}\delta(t)$ approximately when α is small and $\mathrm{L}\{f(t)\} = \frac{\mathrm{E}\alpha}{\mathrm{R}}$.

Therefore $\mathrm{L}\{q\} = \frac{\mathrm{E}\alpha}{\mathrm{R}}\left(\frac{1}{s + \frac{1}{\mathrm{RC}}}\right)$ and $q = \frac{\mathrm{E}\alpha}{\mathrm{R}} e^{-t/RC} u(t)$.

(d) $f(t) = \frac{\mathrm{E}\alpha}{\mathrm{R}}(\delta(t) + \delta(t-\mathrm{T}) + \delta(t-2\mathrm{T}) + \delta(t-3\mathrm{T}) + \ldots)$ approximately for small α

and $\mathrm{L}\{f(t)\} = \frac{\mathrm{E}\alpha}{\mathrm{R}}(1 + e^{-sT} + e^{-s2T} + e^{-s3T} + \ldots)$.

[Alternatively $L\{f(t)\} = \dfrac{1}{1-e^{-sT}} \displaystyle\int_0^T \frac{E\alpha}{R}\,\delta(t)dt$ by Theorem 4.2, since $f(t)$ is periodic.

i.e. $$L\{f(t)\} = \frac{E\alpha/R}{1-e^{-sT}} = \frac{E\alpha}{R}(1 + e^{-sT} + e^{-s2T} + \ldots)\,.]$$

Therefore $$L\{q\} = \frac{E\alpha}{R\left(s+\dfrac{1}{RC}\right)}(1 + e^{-sT} + e^{-s2T} + e^{-s3T} + \ldots)$$

and $$q = \frac{E\alpha}{R}(e^{-t/RC}u(t) + e^{-(t-T)/RC}\,u(t-T) + e^{-(t-2T)/RC}\,u(t-2T) + \ldots)$$

i.e. $$q = \begin{cases} \dfrac{E\alpha}{R}\,e^{-t/RC} & \text{for } 0 \le t < T \\ \dfrac{E\alpha}{R}(e^{-t/RC} + e^{-(t-T)/RC}) & \text{for } T \le t < 2T \\ \dfrac{E\alpha}{R}(e^{-t/RC} + e^{-(t-T)/RC} + e^{-(t-2T)/RC}) & \text{for } 2T \le t < 3T \end{cases}$$

etc.

The graphs of $q \sim t$ for the four cases are shown in Figure 4.7.

Example 4.11

Repeat part (d) of Example 4.10 when E is replaced by

$$E\cos\frac{2\pi}{T}t\,.$$

Now $\dot{q} + \dfrac{q}{RC} = f(t)$ where

$$f(t) = \frac{E\alpha}{R}(\cos 0.\delta(t) + \cos 2\pi.\delta(t-T) + \cos 4\pi.\delta(t-2T) + \ldots) \text{ approximately.}$$

Therefore $$f(t) = \frac{E\alpha}{R}(\delta(t) + \delta(t-T) + \delta(t-2T) + \ldots),$$

which is the *same function* as in Example 4.10(d). Thus the capacitor charge q is identical to that given in Example 4.10(d). The reason for this is that the switch is operating as a *sampling device* on the function $E\cos\frac{2\pi}{T}t$. The only significant values of this function are those at $t = 0, T, 2T, 3T, \ldots$ and all of these are equal to E. Thus as far as the sampling device is concerned the function being sampled is effectively a constant.

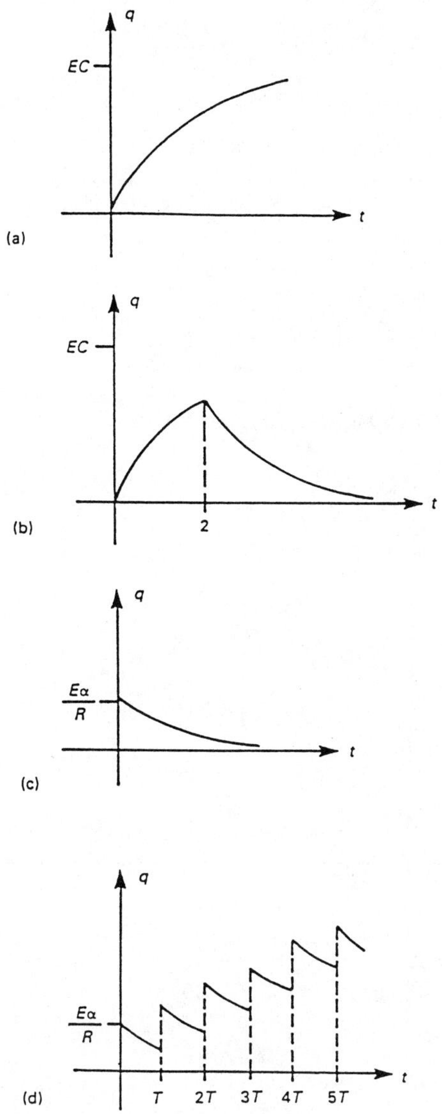

Fig. 4.7

Example 4.12

Find the output of the system $G(s) = \dfrac{1 - e^{-s}}{s}$ when the input is $f(t)$ defined by

$$f(t) = \begin{cases} 2t & \text{when } 0 \leq t < 3 \\ 2 & \text{when } t \geq 3 \end{cases} .$$

Sketch the input and output functions for $0 \leq t < 5$.

$f(t) = 2tu(t) - 2(t-3)\,u(t-3) - 4u(t-3)$, the form required for application of the Shift Theorem, and $L\{f(t)\} = \dfrac{2}{s^2} - \dfrac{2e^{-3s}}{s^2} - \dfrac{4e^{-3s}}{s}$.

Therefore $\quad L\{y(t)\} = \dfrac{2}{s^3}(1-e^{-3s})(1-e^{-s}) - \dfrac{4e^{-3s}}{s^2}(1-e^{-s})$

$$= \frac{2}{s^3}(1-e^{-s}-e^{-3s}+e^{-4s}) - \frac{4}{s^2}(e^{-3s}-e^{-4s})\,.$$

Therefore $\quad y(t) = t^2u(t) - (t-1)^2\,u(t-1) - (t-3)^2\,u(t-3) + (t-4)^2\,u(t-4)$

$$- 4(t-3)\,u(t-3) + 4(t-4)\,u(t-4),$$

i.e., $$y(t) = \begin{cases} t^2 & \text{when } 0 \le t < 1 \\ 2t-1 & \text{when } 1 \le t < 3 \\ -t^2+4t+2 & \text{when } 3 \le t < 4 \\ 2 & \text{when } t \ge 4 \end{cases} .$$

The graphs of $f(t)$ and $y(t)$ are shown in Figure 4.8.

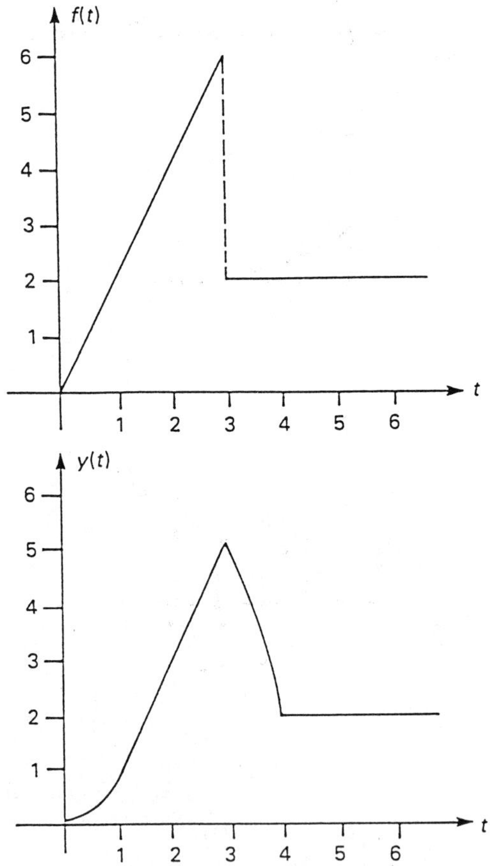

Fig. 4.8

The transfer function considered in this example is a special case of $\dfrac{1-e^{-sT}}{s}$ which is important in the study of sampling techniques and reconstruction of sampled functions. Sampling and reconstruction are considered in Chapter 7.

Problems

1. Solve the differential equation

$$\ddot{y} + 2\dot{y} + y = x(t) \ , \ y(0) = \dot{y}(0) = 0 \ ,$$

where
$$x(t) = \begin{cases} t & \text{when } 0 \le t < 1 \\ 0 & \text{when } t \ge 1 \end{cases} .$$

2. Repeat problem 1 when

(i)
$$x(t) = \begin{cases} 1 & \text{when } 0 \le t < 2 \\ 0 & \text{when } t \ge 2 \end{cases} .$$

(ii)
$$x(t) = \begin{cases} 2 \times 10^5 & \text{when } 3 \le t < 3.001 \\ 0 & \text{otherwise} \end{cases} .$$

3. Solve the differential equation

$$\ddot{x} + 2\dot{x} + 5x = 10\delta(t) \ , \ x(0) = 0 \ , \ \dot{x}(0) = 10 \ .$$

4. An undamped mechanical system with natural frequency w_n radians per second is modelled by the transfer function $G(s) = \dfrac{w_n^2}{s^2 + w_n^2}$. Its input is the periodic square wave $x_p(t)$ defined by

$$x_p(t) = \begin{cases} 1 & \text{when } 0 \le t < L \\ 0 & \text{when } L \le t < 2L \end{cases} , \quad x_p(t + 2L) = x_p(t) \ .$$

Consideration of the Fourier series expansion of $x_p(t)$ shows that we can expect resonance whenever L is such that

$$w_n = \frac{(2r-1)\pi}{L} \ , \quad r = 1, 2, 3, \ldots.$$

Confirm this using Laplace transform techniques.

5. A uniform beam is simply supported at its ends $x = 0$ and $x = L$. The beam has weight W and is subjected to a concentrated load P at $x = \frac{L}{2}$. The deflection of the beam at position x is given by the solution of the differential equation

$$EI\frac{d^4y}{dx^4} = \frac{W}{L} + P\delta\left(x - \frac{L}{2}\right)$$

where E and I are constants. The boundary conditions for the simply supported beam are

$$y(0) = y(L) = 0 \ , \ y''(0) = y''(L) = 0.$$

Use Laplace transform techniques to express y as a function of x .

6. (i) The motion of a railway truck satisfies the differential equation

$$1000\frac{d^2y}{dt^2} + 100\frac{dy}{dt} = x \ , \ y(0) = \frac{dy}{dt}(0) = 0$$

where y denotes the position of the truck at time t and x is the applied force. Write down the transfer function of this system and determine the output y when the input x is given by

$$x = \begin{cases} 1000 & \text{when } t \geq 0 \\ 0 & \text{when } t < 0 \end{cases} .$$

(ii) The truck while at rest is struck by a shunting engine such that the applied force is

$$x = \begin{cases} 5 \times 10^5 \ , & \text{when } 0 \leq t < 10^{-2} \\ 0 & \text{otherwise} \end{cases} .$$

Explain why x can be expressed approximately in terms of the unit impulse function $\delta(t)$ as $x = 5000\ \delta(t)$. For this impulse input determine the output y of the system and write down the initial velocity of the truck and its position when it comes to rest.

What are these values when the impulse approximation is **not** used? Show that $\dot{y}(10^{-2}) \approx 5$.

7. Find the output of the system $G(s) = \frac{1}{1 + sT}$ when the input is

$$x(t) = \begin{cases} 2 - t & \text{when } 0 \leq t < 1 \\ 1 & \text{when } t \geq 1 \end{cases} .$$

8. Find the output of the system $G(s) = \frac{1 - e^{-3s}}{s(s+1)}$ when the input is a unit step function.

9. The input to the filter shown in Figure 4.9 is given by

$$x(t) = \begin{cases} 1 & \text{when } 0 \leq t < 2 \\ 0 & \text{when } t \geq 2 \end{cases} .$$

Find the output $y(t)$ for $t \geq 2$.

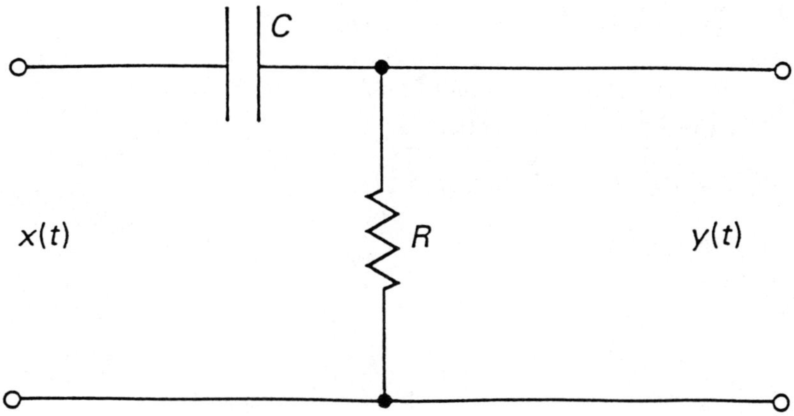

Fig. 4.9

10. In Example 4.10(d) show that

$$q = \frac{E\alpha}{R}\left(\frac{1 - e^{-(n+1)T/RC}}{1 - e^{-T/RC}}\right) \text{ when } t = nT^{+}.$$

What is the maximum level reached by q ?

Answers to problems

1. $$y = \begin{cases} t - 2 + (t+2)e^{-t} & \text{when } 0 \le t < 1 \\ (t+2)e^{-t} - e^{-(t-1)} & \text{when } t \ge 1 \end{cases}.$$

2. (i) $$y = \begin{cases} 1 - (t+1)e^{-t} & \text{when } 0 \le t < 2 \\ -(t+1)e^{-t} + (t-1)e^{-(t-2)} & \text{when } t \ge 2 \end{cases}.$$

(ii) $y = 200(t-3)e^{-(t-3)}\, u(t-3)$.

3. $x = 10e^{-t} \sin 2t\, u(t)$.

5. $$y = \frac{1}{48EI}\left(\frac{2W}{L}x^4 + 8P\left(x - \frac{L}{2}\right)^3 u\left(x - \frac{L}{2}\right) - 4(W+P)x^3 + (2W+3P)L^2x\right),$$
$$0 \le x \le L.$$

6. (i) $\dfrac{1}{1000s(s+0.1)}$, $y = (10t - 100 + 100e^{-0.1t})\,u(t)$.

(ii) $y = (50 - 50e^{-0.1t})\,u(t)$, $\dot{y}(0^+) = 5$, $y(\infty) = 50$.

For *pulse* input $\dot{y}(0^+) = 0$, $y(\infty) = 50$.

7. $$y = \begin{cases} -t + (T+2)(1 - e^{-t/T}) & \text{when } 0 \le t < 1 \\ 1 - (T+2)e^{-t/T} + Te^{-(t-1)/T} & \text{when } t \ge 1 \end{cases}.$$

8. $$y = \begin{cases} t - 1 + e^{-t} & \text{when } 0 \le t < 3 \\ 3 + e^{-t} - e^{-(t-3)} & \text{when } t \ge 3 \end{cases}.$$

9. $y(t) = e^{-t/RC} - e^{-(t-2)/RC}$ when $t \ge 2$.

10. $\dfrac{E\alpha}{R}\left(\dfrac{1}{1 - e^{-T/RC}}\right)$.

PART II: DISCRETE SYSTEMS

5

From the Ideal Sampler to the Z-Transform

5.1 INTRODUCTION

In this chapter we define a theoretical device, *the ideal sampler*, which is similar to the type of sampler used in Example 4.10(d). The output of the ideal sampler is made the input to a continuous system whose output leads to the formulation of the z-transform. The z-transform is a useful tool in the analysis of linear digital systems, that is systems where the input and output are *sequences of numbers* related by the operation of a digital processor. Digital systems will be studied in detail in Chapter 6.

5.2 AN APPROACH TO THE Z-TRANSFORM

The switch sampler described in Example 4.10(d) on page 96 had the effect of replacing the constant $\frac{E}{R}$ by

$$\frac{E\alpha}{R}\left(\delta(t) + \delta(t - T) + \delta(t - 2T) + \ldots\ldots\right),$$

where α was the closure time of the switch (assumed small). If such a sampler is applied to the function $f(t)$ the output would be represented by

$$\alpha\left(f(0)\,\delta(t) + f(T)\,\delta(t - T) + f(2T)\,\delta(t - 2T) + \ldots\ldots\right).$$

This is not a very satisfactory sampling device since the strength of the output signal depends on the closure time of the switch. To overcome this disadvantage it would be necessary to amplify the output signal by an amount $\frac{1}{\alpha}$ so that the output becomes

$$\sum_{n=0}^{\infty} f(nT)\,\delta(t - nT).$$

We define a theoretical device known as the **ideal sampler** which converts the continuous function $f(t)$ into the train of impulses

$$\sum_{n=0}^{\infty} f(nT)\,\delta(t - nT).$$

The notation $f^*(t)$ is used to denote the train of impulses and the symbol used for the ideal sampler is

where T is the sampling interval. Figure 5.1 shows a typical situation.

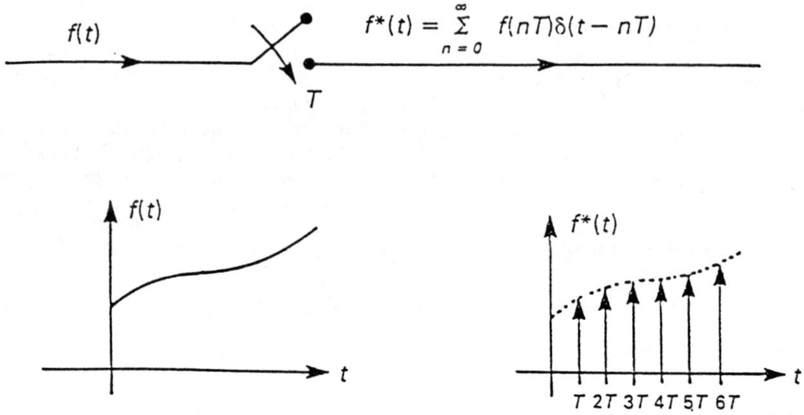

Fig. 5.1 The ideal sampler. The arrow at $t = nT$ represents an impulse of strength $f(nT)$.

The analysis of a system containing an ideal sampler may be carried out using Laplace transforms. Example 5.1 demonstrates the technique.

Example 5.1

Find the output of the sampled-data system shown in Figure 5.2. given that the input $x(t) = t$, $t \geq 0$.

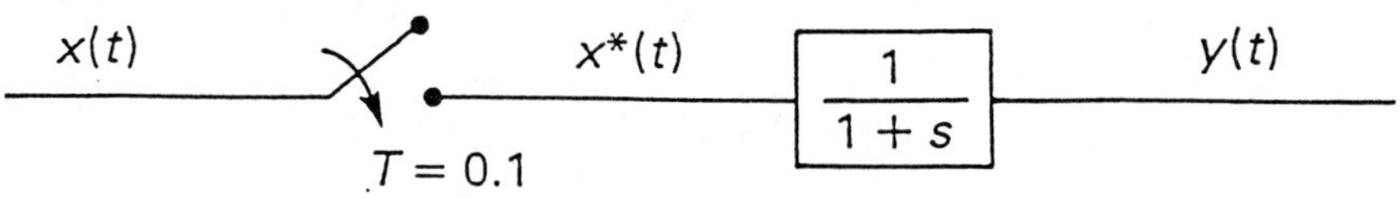

Fig. 5.2

$x(t) = t$ and $x^*(t) = 0.1\delta(t-0.1) + 0.2\delta(t-0.2) + 0.3\delta(t-0.3) + \ldots$.

Therefore $\quad L\{x^*(t)\} = 0.1e^{-0.1s} + 0.2e^{-0.2s} + 0.3e^{-0.3s} + \ldots$

and $\quad L\{y(t)\} = \dfrac{1}{1+s}(0.1e^{-0.1s} + 0.2e^{-0.2s} + 0.3e^{-0.3s} + \ldots)$.

Therefore $y(t) = 0.1e^{-(t-0.1)}u(t-0.1) + 0.2e^{-(t-0.2)}u(t-0.2) + 0.3e^{-(t-0.3)}u(t-0.3) + \ldots$

i.e. $\quad y(t) = 0$ for $0 \le t < 0.1$,

and $\quad y(t) = \sum_{r=1}^{n} 0.1\, r\, e^{-(t-0.1r)}$ for $0.1n \le t < 0.1(n+1)$, $n = 1, 2, 3, \ldots\ldots$.

Suppose that the system considered in Example 5.1 has its output sampled at times 0, 0.1, 0.2, 0.3, ..., to generate the sequence $\{y(0), y(0.1), y(0.2), \ldots\}$ denoted by $\{y(0.1n)\}$ or $\{y_n\}$. y_n can be readily obtained from the result of Example 5.1,

$$y_0 = 0,\ y(0.1n) = \sum_{r=1}^{n} 0.1\, r\, e^{-0.1(n-r)}\ ,\ n = 1, 2, 3, \ldots\ldots \quad (5.1)$$

This process can be generalised for the system shown in Figure 5.3.

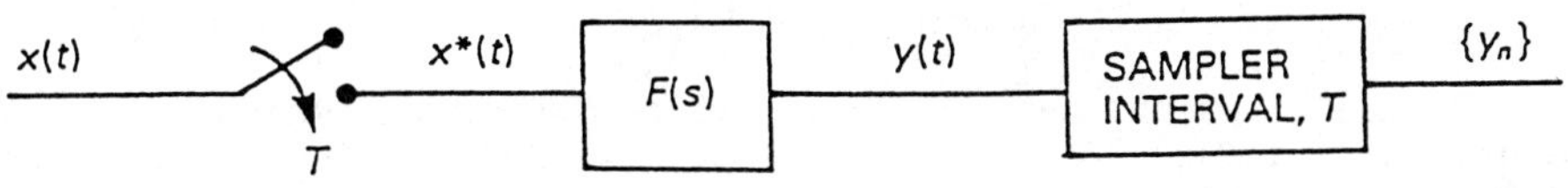

Fig. 5.3

Using Laplace transforms we have

$$L\{y(t)\} = F(s)\, L\{x^*(t)\}$$
$$= F(s) \sum_{r=0}^{\infty} x(rT)e^{-srT} .$$

Therefore

$$y(t) = L^{-1}\left\{F(s) \sum_{r=0}^{\infty} x(rT)e^{-srT}\right\}$$

$$= \sum_{r=0}^{\infty} x(rT)\, f(t-rT)u(t-rT)\ , \quad (5.2)$$

using Theorem 4.1, where $f(t) = L^{-1}\{F(s)\}$.

Finally y_n is obtained from equation (5.2) by replacing t by nT, noting that $u((n-r)\mathrm{T}) = 0$ for $r > n$.

i.e.

$$y_n = y(n\mathrm{T}) = \sum_{r=0}^{n} x(r\mathrm{T})f((n-r)\mathrm{T}) \tag{5.3}$$

[Note the **convolution** form of equation (5.3).]

In situations where *only* the sequence $\{y_n\}$ is required and *not* the function $y(t)$, the following alternative formulation is valuable. It is based on the definition of the **z-transform** of a sequence and leads to *transfer function theory* applicable to digital systems.

We define the z-transform of the sequence of numbers $\{x_n\} = \{x_0, x_1, x_2, x_3, \ldots\}$ as the following infinite series in *descending* powers of z. We denote the z-transform of the sequence by $Z\{x_n\}$.

$$Z\{x_n\} = \sum_{n=0}^{\infty} x_n z^{-n} = x_0 + x_1 z^{-1} + x_2 z^{-2} + \ldots \tag{5.4}$$

The series will be convergent for $|z| > R$ for some number R depending on $\{x_n\}$.

Frequently the sequence $\{x_n\}$ arises from the sampling of a function $x(t)$ at times $t = 0$, T, 2T, 3T, … .

i.e.

$$\{x_n\} = \{x(n\mathrm{T})\}$$

and

$$Z\{x(n\mathrm{T})\} = \sum_{n=0}^{\infty} x(n\mathrm{T})z^{-n}.$$

This is commonly written

$$Z\{x(t)\} = \sum_{n=0}^{\infty} x(n\mathrm{T})z^{-n} = x(0) + x(\mathrm{T})z^{-1} + x(2\mathrm{T})z^{-2} + \ldots \tag{5.5}$$

When writing $Z\{x(t)\}$ we must realise that the z-transformation applies *not* to $x(t)$ but to the sequence $\{x(0), x(\mathrm{T}), x(2\mathrm{T}), x(3\mathrm{T}), \ldots\}$. The z-transform process has the effect of converting a *sequence* into an *infinite series.*

Note that the z-transform of the sequence $\{x(n\mathrm{T})\}$ is equivalent to the Laplace transform of $x^*(t)$ with e^{sT} replaced by z, since

$$L\{x^*(t)\} = \sum_{n=0}^{\infty} x(n\mathrm{T})e^{-snT}, \quad \text{and}$$

$$Z\{x(t)\} = \sum_{n=0}^{\infty} x(n\mathrm{T})z^{-n}.$$

The relationship $z = e^{sT}$ is used extensively throughout the remaining chapters.

Returning to the system of Figure 5.3 and noting its convolution relationship (5.3), we state a *transfer function relationship* in terms of z-transforms. (We have already seen in Chapter 1 the correspondence between the convolution integral and transfer function for a continuous system.)

The transfer function form corresponding to Figure 5.3 is

$$Z\{y(t)\} = Z\{f(t)\} \,.\, Z\{x(t)\} \tag{5.6}$$

where

$$f(t) = L^{-1}\{F(s)\} \tag{5.7}$$

$Z\{f(t)\}$ is the transfer function of the system relating the input function $x(t)$ to the output *sequence* $\{y(n\mathrm{T})\}$. Denoting the transfer function by $F(z)$ the required output sequence is obtained by taking the *inverse* z-transform

$$\{y(n\mathrm{T})\} = Z^{-1}\{F(z).\, Z\{x(t)\}\} \tag{5.8}$$

The inverse z-transform converts an infinite series in z^{-1} to a sequence of numbers as follows

$$Z^{-1}\{y_0 + z^{-1}y_1 + z^{-2}y_2 + \ldots\ldots\} = \{y_n\} \tag{5.9}$$

It is important to understand that the *continuous function y(t)* cannot be obtained by equation (5.8).

The formulation (5.8) is demonstrated in Example 5.2.

Example 5.2

Write down the transfer function corresponding to Figure 5.2. Use z-transforms to obtain the output sequence $\{y(0.1n)\}$ for $x(t) = t$.

$F(s) = \dfrac{1}{1+s}$, therefore $f(t) = e^{-t}$.

It follows that

$$F(z) = 1 + e^{-0.1}\,z^{-1} + e^{-0.2}z^{-2} + e^{-0.3}z^{-3} + \ldots\ldots$$

is the transfer function relating $x(t)$ and $\{y(0.1n)\}$.

Now $x(t) = t$ and therefore

$$Z\{x(t)\} = 0.1z^{-1} + 0.2z^{-2} + 0.3z^{-3} + \ldots\ldots.$$

From equation (5.8) the z-transform of $\{y(0.1n)\}$ is obtained by taking the product of these infinite series.

Therefore

$$Z\{y(0.1n)\} = 0.1z^{-1} + (0.2 + 0.1e^{-0.1})z^{-2} + (0.3 + 0.2e^{-0.1} + 0.1e^{-0.2})z^{-3} + \ldots\ldots.$$

The output sequence $\{y(0.1n)\}$ is obtained by taking the inverse z-transform, i.e. the set of coefficients of descending powers of z.

Therefore

$$\begin{aligned} y(0) &= 0 \quad \text{(the coefficient of } z^{0}\text{)},\\ y(0.1) &= 0.1 \quad \text{(the coefficient of } z^{-1}\text{)},\\ y(0.2) &= 0.2 + 0.1e^{-0.1} \quad \text{(the coefficient of } z^{-2}\text{)},\\ y(0.3) &= 0.3 + 0.2e^{-0.1} + 0.1e^{-0.2} \quad \text{(the coefficient of } z^{-3}\text{)}. \end{aligned}$$

It is easily seen that $y(0) = 0$

and $$y(0.1n) = \sum_{r=1}^{n} 0.1re^{-0.1(n-r)}\ , \quad n = 1, 2, 3, \ldots$$

consistent with the result obtained in equation (5.1).

The above example demonstrates that an inverse z-transform can be obtained simply by taking the coefficients in an infinite series. Other methods for obtaining inverse z-transforms are considered in Section 6.4.

Problems

1. Obtain the output $y(t)$ of the system shown in Figure 5.4 when the input $x(t)$ is the unit step function.

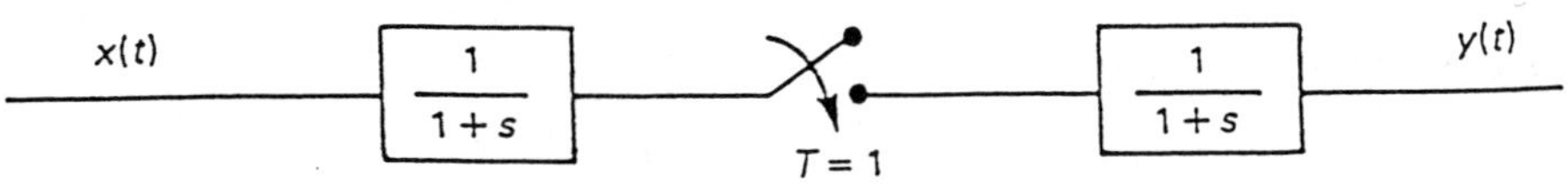

Fig. 5.4

2. Repeat Example 5.2 for $x(t) = 1 - e^{-t}$ and T = 1. Compare your answer with Problem 1.

Answers to problems

1. $$y(t) = \frac{e(e^n - 1)e^{-t}}{e-1} - ne^{-t}\ , \quad n \le t < n+1, \quad n = 0, 1, 2, \ldots .$$

5.3 Z-TRANSFORMS OF SIMPLE SEQUENCES

For many simple sequences the infinite series can be summed to obtain the z-transform in closed form. It is worth remembering the particular series

$$\frac{1}{1-x} = 1 + x + x^2 + x^3 + \ldots, \ |x| < 1.$$

$$\frac{1}{(1-x)^2} = 1 + 2x + 3x^2 + 4x^3 + \ldots, \ |x| < 1.$$

The following examples illustrate the technique.

(i) The sequence $x_n = 1, n = 0, 1, 2, 3, \ldots$, i.e. $\{x_n\} = \{1, 1, 1, \ldots\}$.

$$Z\{x_n\} = 1 + z^{-1} + z^{-2} + z^{-3} + \ldots\ldots$$

$$= \frac{1}{1 - z^{-1}}$$

$$= \frac{z}{z - 1} \; .$$

Alternatively we can write $Z\{u(t)\} = \frac{z}{z-1}$, where $u(t)$ is the unit step function. Note that $Z\{u(t)\}$ means $Z\{u(nT)\} = Z\{1, 1, 1, \ldots\}$.

(ii) The sequence $\{x_n\} = \{c^n\} = \{1, c, c^2, c^3, \ldots\ldots\}$ where c is a constant.

$$Z\{x_n\} = 1 + cz^{-1} + c^2z^{-2} + c^3z^{-3} + \ldots\ldots$$

$$= \frac{1}{1 - cz^{-1}}$$

$$= \frac{z}{z - c} \; .$$

(iii) If we sample the function $x(t) = a^t$ at times $t = 0, T, 2T, 3T, \ldots\ldots$, we obtain the sequence $\{a^{nT}\}$ which may be written $\{(a^T)^n\}$. Using the result of (ii) we have

$$Z\{a^{nT}\} = \frac{z}{z - a^T}$$

which may be written

$$Z\{a^t\} = \frac{z}{z - a^T} \; .$$

(iv) The sequence $\{x_n\} = \{n\} = \{0, 1, 2, 3, \ldots\}$.

$$Z\{x_n\} = z^{-1} + 2z^{-2} + 3z^{-3} + 4z^{-4} + \ldots\ldots$$

$$= z^{-1}(1 + 2z^{-1} + 3z^{-2} + 4z^{-3} + \ldots\ldots)$$

$$= z^{-1}(1 - z^{-1})^{-2}$$

$$= \frac{z}{(z - 1)^2} \; .$$

(v) To obtain the z-transform of the sequence obtained by sampling $x(t) = t$, $t = 0$, T, 2T, 3T, ..., i.e. the sequence $\{nT\}$, we note that $\{nT\} = T\{n\}$ and from (iv) we have $Z\{t\} = \frac{Tz}{(z-1)^2}$.

Table 2A gives the z-transforms of common sequences $\{x_n\}$ where x_n is given as a function of the integral variable n. This table may be used to obtain directly the results (i), (ii) and (iv).

TABLE 2A

Table of z-transforms of $\{x_n\}$, $x_n = 0$ for $n < 0$

$$Z\{x_n\} = \sum_{n=0}^{\infty} x_n z^{-n} = F(z)$$

x_n, $n = 0, 1, 2, \ldots$	$F(z)$
1	$\dfrac{z}{z-1}$
n	$\dfrac{z}{(z-1)^2}$
n^2	$\dfrac{z(z+1)}{(z-1)^3}$
$n\ x_n$	$-zF'(z)$
$u(n-1)$	$\dfrac{1}{z-1}$
$u(n-k)$, $k = 1,2,3,\ldots$	$\dfrac{1}{z^{k-1}(z-1)}$
$\delta_n = u(n) - u(n-1)$	1
$\delta_{n-k} = u(n-k) - u(n-k-1)$	$\dfrac{1}{z^k}$
a^n	$\dfrac{z}{z-a}$
e^{cn}	$\dfrac{z}{z-e^c}$
$n\ a^{n-1}$	$\dfrac{z}{(z-a)^2}$
$a^{n-1} u(n-1)$	$\dfrac{1}{z-a}$
$\cos bn$	$\dfrac{z(z-\cos b)}{z^2 - 2z\cos b + 1}$

$\sin bn$	$\dfrac{z \sin b}{z^2 - 2z \cos b + 1}$
$a^n \, x_n$	$F\,(a^{-1} z)$
x_{n+1}	$z\,F(z) - z\,x_0$
x_{n+2}	$z^2\,F(z) - z^2\,x_0 - z\,x_1$
x_{n+k}	$z^k\,F(z) - z^k \sum_{i=0}^{k-1} x_i\,z^{-i}$
$x_{n-k}\,u(n-k)$	$z^{-k}\,F(z)$

Table 2B gives the z-transforms of sequences formed by sampling certain common functions of t at times $t = 0$, T, 2T, 3T, ..., and may be used to obtain the results (iii) and (v).

The properties of the z-transform given on pages 115-116 follow directly from the definition

$$\mathrm{Z}\{x_n\} = \sum_{n=0}^{\infty} x_n z^{-n} .$$

The properties are useful for obtaining the z–transforms of non-trivial functions, and for dealing with difference equations (see Chapter 6).

All of the properties correspond to results already obtained for the Laplace transform. In particular property (ii) demonstrates that multiplication by z^{-1} has the effect of *delaying* the sequence by one sampling interval. This should not be surprising since

$$z^{-1} = (e^{sT})^{-1} = e^{-sT}$$

and we have already seen that, for continuous functions, multiplication by e^{-sT} is equivalent to a delay of amount T, (Theorem 4.1).

For properties (ii) and (iv) it is necessary to adopt the following convention.

For all sequences $\{x_n\}$ we define $x_n = 0$ when $n < 0$. This is reasonable for systems where x_0 is the "first" number entering (usually at time $t = 0$). Since there are no inputs before this time it is clear that $x_{-1} = x_{-2} = \ldots\ldots = 0$. It follows that the sequence $\{x_{n-1}\} = \{x_{-1}, x_0, x_1, x_2, \ldots\ldots\} = \{0, x_0, x_1, x_2, \ldots\}$. $\{x_{n-1}\}$ is the sequence $\{x_n\}$ *delayed* by one sampling interval, referred to as the *delayed sequence*. Similarly we have $\{x_{n-2}\} = \{0, 0, x_0, x_1, x_2, \ldots\}$, the sequence $\{x_n\}$ delayed by two sampling intervals.

Property (iii) is proved in Example 5.4. Proof of the other properties is left for the reader.

TABLE 2B

Table of z-transforms of $f(t)$, $f(t) = 0$ for $t < 0$

$$Z\{f(t)\} = \sum_{n=0}^{\infty} f(nT) z^{-n} = F(z)$$

$f(t)$, $t = 0, T, 2T, \ldots$	$F(z)$
1	$\dfrac{z}{z-1}$
t	$\dfrac{zT}{(z-1)^2}$
t^2	$\dfrac{T^2 z(z+1)}{(z-1)^3}$
$t\,f(t)$	$-zTF'(z)$
$u(t-T)$	$\dfrac{1}{z-1}$
$u(t) - u(t-T)$	1
$u(t-kT)$, $k = 1,2,3,\ldots$	$\dfrac{1}{z^{k-1}(z-1)}$
$u(t-kT) - u(t-[k+1]T)$	$\dfrac{1}{z^k}$
a^t	$\dfrac{z}{z-a^T}$
e^{ct}	$\dfrac{z}{z-e^{cT}}$
$a^{t/T}$	$\dfrac{z}{z-a}$
$\dfrac{t}{T} a^{(t-T)/T}$	$\dfrac{z}{(z-a)^2}$
$a^{(t-T)/T} u(t-T)$	$\dfrac{1}{z-a}$
$\cos bt$	$\dfrac{z(z-\cos bT)}{z^2-2z\cos bT+1}$

$\sin bt$	$\dfrac{z \sin bT}{z^2 - 2z \cos bT + 1}$
$e^{ct} f(t)$	$F(e^{-cT} z)$
$f(t+T)$	$zF(z) - z f(0)$
$f(t+2T)$	$z^2 F(z) - z^2 f(0) - z f(T)$
$f(t+kT)$	$z^k F(z) - z^k \sum_{i=0}^{k-1} f(iT) z^{-i}$
$f(t-kT)\, u(t-kT)$	$z^{-k} F(z)$

Convention for Jump Discontinuities

If $f(t)$ has a jump discontinuity at $t = nT$, we interpret $f(nT)$ as follows:

$$f(nT) = \lim_{t \to nT+} f(t)$$

Further properties of z-transforms of the sequence $\{x_n\}$.

(i) (Linearity property)

$Z\{c_1 x_n + c_2 y_n\} = c_1 Z\{x_n\} + c_2 Z\{y_n\}$.

(ii) (Delay Operation)

$Z\{x_{n-1}\} = z^{-1} Z\{x_n\}$. This is the *shift theorem* for z-transforms.

(iii) $Z\{x_{n+1}\} = z\, Z\{x_n\} - z\, x(0)$.

(iv) $Z\{x_{n-k}\} = z^{-k} Z\{x_n\}$, $k = 1, 2, 3, \ldots \ldots\ldots$.

(v) $Z\{x_{n+k}\} = z^k Z\{x_n\} - \sum_{i=0}^{k-1} z^{k-i} x_i$, $k = 1, 2, 3, \ldots$.

(vi) $Z\{a^n x_n\} = F(a^{-1} z)$ where $F(z) = Z\{x_n\}$.

Example (Note that the sequence $\{1, 1, 1, \ldots\ldots\}$ is written $\{1\}$.

$$Z\{1\} = \frac{z}{z-1}.$$

Therefore $\quad Z\{a^n . 1\} = \dfrac{a^{-1}z}{a^{-1}z - 1}$,

i.e. $\quad Z\{a^n\} = \dfrac{z}{z-a}$.

(vii) $\quad Z\{n x_n\} = -z F'(z)$ where $F(z) = Z\{x_n\}$.

Example

$$Z\{n\} = \frac{z}{(z-1)^2} .$$

Therefore
$$Z\{n^2\} = -z\frac{d}{dz}\left(\frac{z}{(z-1)^2}\right)$$

$$= -z\left(\frac{(z-1)^2 - 2z(z-1)}{(z-1)^4}\right)$$

$$= \frac{z(z+1)}{(z-1)^3} .$$

(viii) (Initial Value Theorem)

$$x_0 = \lim_{z\to\infty} (F(z)) .$$

(ix) (Final Value Theorem)

$$\lim_{n\to\infty} (x_n) = \lim_{z\to 1^+} ((z-1)\,F(z)) ,$$

provided the series F(z) is convergent for $|z| > 1$.

Example 5.3

(a) Use the definition of the z-transform to obtain

$$Z\{t\,e^{3t}\} ,\ t = 0, T, 2T, 3T, \ldots\ldots\ldots .$$

(b) Find the sequence $\{x_n\}$ such that

$$Z\{x_n\} = \frac{2z^2}{z^2-9} .$$

(a)
$$Z\{t\,e^{3t}\} = Te^{3T} z^{-1} + 2Te^{6T} z^{-2} + 3Te^{9T} z^{-3} + \ldots\ldots$$

$$= Te^{3T} z^{-1} (1 + 2e^{3T}z^{-1} + 3e^{6T}z^{-2} + \ldots\ldots)$$

$$= Te^{3T} z^{-1} (1 - e^{3T}z^{-1})^{-2}$$

$$= \frac{Te^{3T}z}{(z-e^{3T})^2} .$$

[The above result can be confirmed using Table 2B.

$$Z\{t\} = \frac{Tz}{(z-1)^2} .$$

Therefore
$$Z\{e^{3t}\,t\} = \frac{Te^{-3T}z}{(e^{-3T}z-1)^2} = \frac{Te^{3T}z}{(z-e^{3T})^2} .]$$

(b) Expressing the function as the sum of partial functions *after* taking out z as a factor we obtain

$$\frac{2z^2}{z^2-9} = z\left(\frac{1}{z-3}+\frac{1}{z+3}\right)$$

$$= \left(\frac{z}{z-3}+\frac{z}{z+3}\right).$$

The removal of the factor z in this case allows the partial function method to be applied to an "improper" rational function of z. However, it should be noted that most of the entries in the right-hand column of Table 2A have a factor z in the numerator and it is convenient to arrange this to happen before the table is used to determine inverse z-transforms.

Therefore from Table 2A

$$Z^{-1}\left\{\frac{2z^2}{z^2-9}\right\} = \{3^n + (-3)^n\}$$

and the sequence $\{x_n\}$ is given by

$$\{x_n\} = \{2, 0, 18, 0, 162, \ldots\ldots\}.$$

Example 5.4

Prove property (iii) on page 115.

Property (iii) is

$$Z\{x_{n+1}\} = z\ Z\{x_n\} - z\,x_0, \quad \text{where } \{x_{n+1}\} = \{x_1, x_2, x_3, \ldots\ldots\}.$$

$$Z\{x_{n+1}\} = x_1 + x_2z^{-1} + x_3z^{-2} + \ldots\ldots$$

$$= z(x_0 + x_1z^{-1} + x_2z^{-2} + x_3z^{-3} + \ldots\ldots) - z\,x_0$$

$$= z\ Z\{x_n\} - z\,x_0.$$

Problems

1. Obtain the z-transform of the following sequences.

(i) $\{x_n\} = \{1, 0, 1, 0, 0, 0, 0, \ldots\}$
(ii) $\{x_n\} = \{n+1\}$
(iii) $\{x_n\} = \{1, 0, 1, 0, 1, \ldots\ldots\}$
(iv) $\{x_n\} = \{0, 0, 1, 1, 1, 1, \ldots\}$
(v) $\{u(n-2)\}$
(vi) $\{\sin 3n\}$
(vii) $\{4^n\}$.

2. Determine the inverse z-transform of the following

(i) $4z^{-1}$ (ii) $z^{-1} + z^{-3}$ (iii) $\dfrac{z}{z+1}$

(iv) $\dfrac{1}{z+1}$ (v) $\dfrac{z^{-1}}{(z+1)}$ (vi) $\dfrac{2z-3}{z-1}$

(vii) $\dfrac{z}{z^2-3z+2}$ (viii) $\dfrac{z^2+2z}{z^2-4z+4}$ (ix) $\dfrac{z}{z^2+1}$.

3. Find the sampled sequence $f(t)$, $t = 0$, T, 2T, 3T, ..., such that

(i) $Z\{f(t)\} = \dfrac{3z^3 - 8z^2 + 9z}{(z-2)(z+3)(z-1)}$

(ii) $Z\{f(t)\} = \dfrac{3z^3 - 6z^2 - 5z}{(z-1)^2(z+1)}$.

For (i) and (ii) find the sequence $\{x_n\}$ which has the same z-transform.

4. Use the definition of a z-transform to prove that

$$Z\{(n-1)u(n-1)\} = \frac{1}{(z-1)^2} .$$

The above result is a special case of property (ii) on page 115. Prove this property.

5. Prove properties (iv) and (v) on page 115.

6. Use Table 2A to write down the z-transform of the sequence $\{e^{jbn}\}$.

Hence obtain the z-transforms of cos bn and sin bn in the forms given in Table 2A.

7. Prove that

$$Z\{e^{ct} f(t)\} = F(e^{-cT}z).$$

Hence find the z-transforms of $e^{ct}\cos bt$ and $e^{ct}\sin bt$, $t = 0$, T, 2T, 3T, ...

8. Prove properties (vii) and (viii) on pages 115-116 .

Answers

1. (i) $1 + z^{-2}$ (ii) $\dfrac{z^2}{(z-1)^2}$ (iii) $\dfrac{z^2}{z^2-1}$ (iv) $\dfrac{z^{-1}}{z-1}$

(v) $\dfrac{z^{-1}}{z-1}$ (vi) $\dfrac{z \sin 3}{z^2 - 2z\cos 3 + 1}$ (vii) $\dfrac{z}{z-4}$.

2. (i) $\{0, 4, 0, 0, 0, \dots\}$ (ii) $\{0, 1, 0, 1, 0, 0, 0, \dots\}$ (iii) $\{(-1)^n\}$

(iv) $\{(-1)^{n-1} . u(n-1)\}$ (v) $\{(-1)^n u(n-2)\}$

(vi) $\{2 - 3u(n-1)\}$ (vii) $\{2^n - 1\}$

(viii) $\{2^n + 4n\, 2^{n-1}\}$ (ix) $\left\{\sin \frac{\pi}{2} n\right\} = \{0, 1, 0, -1, 0, \dots\dots\}$.

3. (i) $f(t) = 2^{t/T} + 3(-3)^{t/T} - 1\ ,\ t = 0, T, 2T, 3T, \dots.$

$$\{x_n\} = \{2^n + 3(-3)^n - 1\}\ .$$

(ii) $f(t) = 2 - 4\left(\frac{t}{T}\right) + (-1)^{t/T}\ , t = 0, T, 2T, 3T, \dots.,$

$$\{x_n\} = \{2 - 4n + (-1)^n\}\ .$$

7. $$Z\{e^{ct} \cos bt\} = \frac{z^2 - ze^{cT} \cos bT}{z^2 - 2ze^{cT} \cos bT + e^{2cT}}$$

$$Z\{e^{ct} \sin bt\} = \frac{ze^{cT} \sin bT}{z^2 - 2ze^{cT} \cos bT + e^{2cT}}\ .$$

6

Solution of Linear Difference Equations

6.1 INTRODUCTION

The concept of a Digital System was introduced in Section 5.2. When such a system is *linear* the output sequence $\{y_n\}$ and input sequence $\{x_n\}$ may be related by a linear difference equation.

In this chapter z-transform techniques are applied to the solution of linear difference equations. For simple cases z-transforms are not necessarily the best method of solution but the corresponding transfer function notation is useful in subsequent work. Attention will be drawn to the similarities between the transfer function approaches to differential equations and difference equations. In particular the frequency characteristics of discrete and continuous systems will be compared.

6.2 DIFFERENCE EQUATIONS

It is convenient to think of the elements of the input sequence $\{x_n\}$ of a digital system to enter the system at specified equally spaced times 0, T, 2T, ..., i.e. x_0 enters the system at time $t = 0$, x_1 at time $t = \mathrm{T}$, x_2 at time $t = 2\mathrm{T}$, ... Similarly, if $\{y_n\}$ is the output sequence, y_0 leaves the system at $t = 0$, y_1 at $t = \mathrm{T}$, y_2 at $t = 2\mathrm{T}$, ... It is assumed that the time taken to calculate a specific output element is negligible compared with T.

A difference equation expresses a particular output y_n of a digital system in terms of the synchronous input x_n and earlier outputs and inputs. For a *linear* digital system we have

$$y_n = \sum_{i=1}^{N} a_i y_{n-i} + \sum_{i=0}^{M} b_i x_{n-i}, \quad a_N \neq 0, \; b_M \neq 0 \tag{6.1}$$

The integer N is the **order** of the difference equation. It is assumed that for $n < 0$, $x_n = y_n = 0$.

We recall the definition of the z-transform given in Chapter 5. The z-transform converts a sequence of numbers $\{x_n\} = \{x_0, x_1, x_2, \ldots\}$ into an infinite series in descending powers of z,

$$Z\{x_n\} = x_0 + x_1 z^{-1} + x_2 z^{-2} + \ldots\ldots\ldots \;.$$

The inverse z-transform reverses this operation, constructing a sequence from the coefficients in an infinite series.

Using the delay property $Z\{x_{n-k}\} = z^{-k} Z\{x_n\}$(property (iv) in Section 5.3), we can transform equation (6.1) to obtain

$$Z\{y_n\} = \sum_{i=1}^{N} a_i \, z^{-i} \, Z\{y_n\} + \sum_{i=0}^{M} b_i \, z^{-i} \, Z\{x_n\}$$

i.e.

$$Z\{y_n\} = \left(\sum_{i=0}^{M} b_i \, z^{-i} \Big/ \left(1 - \sum_{i=1}^{N} a_i \, z^{-i} \right) \right) Z\{x_n\}$$

i.e. $$Z\{y_n\} = H(z) \,.\, Z\{x_n\} ,$$

where $$H(z) = \sum_{i=0}^{M} b_i \, z^{N-i} \Big/ \left(z^N - \sum_{i=1}^{N} a_i \, z^{N-i} \right)$$

is called the **transfer function** of the digital system.

We may now use **block diagram notation** for digital systems as shown in figure 6.1.

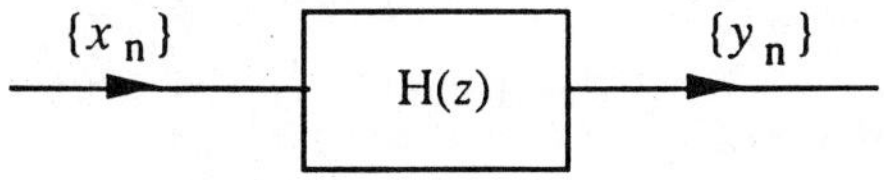

Fig. 6.1 Block diagram representation of digital system with transfer function $H(z)$, that is, $Z\{y_n\} = Z\{x_n\} \, H(z)$.

The worked examples show the use of z-transforms in the solution of difference equations.

Three particular input sequences are worth mentioning.

(i) the **unit step** sequence $\{u_n\}$ where $u_n = 1, n = 0, 1, 2, 3, \ldots,$

$$Z\{u_n\} = \frac{z}{z-1} \,.$$

(ii) The **unit impulse** sequence $\{\delta_n\}$ where $\delta_n = \begin{cases} 1 , & n = 0 \\ 0 , & n = 1, 2, 3, \ldots \end{cases} ,$

$$Z\{\delta_n\} = 1.$$

(iii) The **delayed unit impulse** $\{\delta_{n-k}\}$ where $\delta_{n-k} = \begin{cases} 0 , & n = 1, 2, 3, \ldots k-1 \\ 1 , & n = k \\ 0 , & n = k+1, k+2, \ldots \end{cases}$

$$Z\{\delta_{n-k}\} = z^{-k} \,.$$

Note that the output of a digital system with transfer function $H(z)$ for a unit impulse input is given by

$$Z\{y_n\} = H(z) \,.\, 1$$

i.e. $y_n = h_n$ where $\{h_n\} = Z^{-1}\{H(z)\}$.

The sequence $\{h_n\}$ is called the **impulse response sequence**.

When the input is $\{\delta_{n-k}\}$ the digital system will have output

$$\{y_n\} = Z^{-1}\{z^{-k} \, H(z)\} = \{h_{n-k}\} ,$$

(where $h_i = 0$ for $i < 0$).

Further since any general input sequence $\{x_n\}$ can be expressed as the infinite sum

$$\{x_n\} = x_0\{\delta_n\} + x_1\{\delta_{n-1}\} + x_2\{\delta_{n-2}\} + \ldots\ldots ,$$

then the output of the system H(z) for input $\{x_n\}$ is given by

$$\{y_n\} = x_0\{h_n\} + x_1\{h_{n-1}\} + x_2\{h_{n-2}\} + \ldots\ldots .$$

i.e.
$$y_n = \sum_{i=0}^{n} x_i h_{n-i} , \quad n = 0, 1, 2, 3, \ldots .$$

This is the *convolution property* for discrete systems which closely resembles the corresponding property for continuous systems given in Section 1.6.

Example 6.1

Solve the difference equation

$$y_n = 3y_{n-2} - 2y_{n-1} + x_n + x_{n-1} , \quad n = 0, 1, 2, 3, \ldots ,$$

given that $\{x_n\}$ is the unit step sequence,

i.e. $\{x_n\} = \{1, 1, 1, 1, \ldots\ldots\}$, and $x_n = y_n = 0$ when $n < 0$.

We can of course obtain the numerical values $y_0, y_1, y_2, \ldots\ldots$ simply by substituting values of n in the given equation.

$$y_0 = 3y_{-2} - 2y_{-1} + x_0 + x_{-1} = 0 - 0 + 1 + 0 = 1,$$

$$y_1 = 3y_{-1} - 2y_0 + x_1 + x_0 = 0 - 2 + 1 + 1 = 0 ,$$

$$y_2 = 3y_0 - 2y_1 + x_2 + x_1 = 3 - 0 + 1 + 1 = 5,$$

$$y_3 = 3y_1 - 2y_2 + x_3 + x_2 = 0 - 10 + 1 + 1 = - 8,$$

$$y_4 = 3y_2 - 2y_3 + x_4 + x_3 = 15 + 16 + 1 + 1 = 33, \quad \text{and so on.}$$

Therefore $\{y_n\} = \{1, 0, 5, -8, 33, \ldots\ldots\}$.

However, it is not easy to establish a *general formula* for y_n from the above sequence.

Taking z-transforms of the difference equation, and using property (iv) in Section 5.3 we obtain

$$Z\{y_n\} = 3z^{-2} Z\{y_n\} - 2z^{-1} Z\{y_n\} + Z\{x_n\} + z^{-1} Z\{x_n\}$$

i.e.
$$Z\{y_n\} = \left(\frac{1 + z^{-1}}{1 + 2z^{-1} - 3z^{-1}}\right) Z\{x_n\}$$

$$= \left(\frac{z^2 + z}{z^2 + 2z - 3}\right)\left(\frac{z}{z - 1}\right) \text{ since } \{x_n\} = \{1\}.$$

Therefore
$$Z\{y_n\} = z\left(\frac{z^2 + z}{(z - 1)^2(z + 3)}\right)$$

$$= \frac{5}{8}\left(\frac{z}{z-1}\right) + \frac{1}{2}\left(\frac{z}{(z-1)^2}\right) + \frac{3}{8}\left(\frac{z}{z+3}\right), \text{ (partial fractions)}.$$

Table 2A is now used to obtain the inverse z-transform and $\{y_n\} = \left\{\frac{5}{8} + \frac{n}{2} + \frac{3}{8}(-3)^n\right\}$ is the general form of the solution of the difference equation.

Example 6.2

Solve the difference equation

$$y_{n+2} + 3y_{n+1} + 2y_n = 0\ , \quad n = 0, 1, 2, 3, \ldots\ldots \text{ given that } y_0 = 1, y_1 = 2.$$

For this type of difference equation we may apply property (v) in Section 5.3.

$$\begin{aligned} Z\{y_{n+2}\} &= z^2 Z\{y_n\} - (z^2 y_0 + z y_1) \\ &= z^2 Z\{y_n\} - z^2 - 2z\ , \\ Z\{y_{n+1}\} &= z\, Z\{y_n\} - z y_0 \\ &= z\, Z\{y_n\} - z. \end{aligned}$$

Transforming the given difference equation gives

$$z^2\, Z\{y_n\} - z^2 - 2z + 3(z\, Z\{y_n\} - z) + 2\, Z\{y_n\} = 0.$$

Therefore $$Z\{y_n\} = \frac{z^2 + 5z}{z^2 + 3z + 2} = \frac{z(z+5)}{(z+1)(z+2)}\ .$$

Applying partial fractions but retaining the factor z we obtain

$$\begin{aligned} Z\{y_n\} &= z\left(\frac{4}{z+1} - \frac{3}{z+2}\right) \\ &= \frac{4z}{z-(-1)} - \frac{3z}{z-(-2)}\ . \end{aligned}$$

Using Table 2A

$$\{y_n\} = \{4(-1)^n - 3(-2)^n\}.$$

The reader will note that the difference equation from Example 6.2 was *not* in the form of equation (6.1), and property (v) was required instead of property (iv). However, the equation may be written as

$$y_n + 3y_{n-1} + 2y_{n-2} = x_n\ ,\ n = 0, 1, 2, 3, \ldots\ ,\ y_{-1} = y_{-2} = 0\ ,$$

where $\{x_n\} = \{1, 5, 0, 0, 0, \ldots\ldots\}$.

This corresponds to the form of equation (6.1), $\{x_n\}$ being chosen to make $y_0 = 1$ and $y_1 = 2$. Now using property (iv) we have

$$(1 + 3z^{-1} + 2z^{-2})\, Z\{y_n\} = 1 + 5z^{-1}\ ,$$

since $Z\{x_n\} = 1 + 5z^{-1}$ directly from the definition of the z-transform. Therefore

$$Z\{y_n\} = \frac{z^2 + 5z}{z^2 + 3z + 2}$$

and the solution continues as above.

Example 6.3

Solve the difference equation

$$y_{n+1} + 2y_n = x_n\ ,\quad n = 0, 1, 2, 3, \ldots\ ,\quad \text{where } \{x_n\} = \{n\} \text{ and } y_0 = 1.$$

First write the equation in the form of equation (6.1).

The given equation is equivalent to

$$y_n + 2y_{n-1} = x_{n-1} + \delta_n\ ,\quad n = 0, 1, 2, 3, \ldots ,$$

where $\{\delta_n\} = \{1, 0, 0, 0, \ldots\}$ is added to ensure that $y_0 = 1$, $(y_{-1} = x_{-1} = 0)$.

Taking z-transforms

$$(1 + 2z^{-1})\, Z\{y_n\} = z^{-1} \frac{z}{(z-1)^2} + 1$$

i.e.

$$\begin{aligned} Z\{y_n\} &= \left(\frac{z}{z+2}\right)\left(\frac{1 + (z-1)^2}{(z-1)^2}\right) \\ &= z\left(\frac{z^2 - 2z + 2}{(z+2)(z-1)^2}\right) \\ &= \frac{1}{3}\left(\frac{z}{(z-1)^2}\right) - \frac{1}{9}\left(\frac{z}{z-1}\right) + \frac{10}{9}\left(\frac{z}{z+2}\right) \end{aligned}$$

and from Table 2A

$$\{y_n\} = \left\{\frac{n}{3} - \frac{1}{9} + \frac{10}{9}(-2)^n\right\}.$$

Example 6.4

A digital system is described by the transfer function $\dfrac{z^2 + 3z}{z^2 + 4z - 21}$. Find the output sequence for a unit step input sequence.

$$Z\{u_n\} = \frac{z}{(z-1)}\ .$$

Therefore

$$\begin{aligned} Z\{y_n\} &= \left(\frac{z^2 + 3z}{z^2 + 4z - 21}\right) \cdot \left(\frac{z}{z-1}\right) \\ &= \frac{z(z^2 + 3z)}{(z-1)(z+7)(z-3)} \\ &= -\frac{1}{4} \cdot \frac{z}{(z-1)} + \frac{7}{20} \cdot \frac{z}{(z+7)} + \frac{9}{10} \cdot \frac{z}{(z-3)}\ . \end{aligned}$$

Therefore

$$\{y_n\} = \left\{-\frac{1}{4} + \frac{7}{20}(-7)^n + \frac{9}{10} \cdot 3^n\right\}.$$

Alternatively we can obtain the first few numerical values by writing down the corresponding difference equation as follows:

$$\frac{Z\{y_n\}}{Z\{x_n\}} = \frac{z^2 + 3z}{z^2 + 4z - 21} = \frac{1 + 3z^{-1}}{1 + 4z^{-1} - 21z^{-2}} .$$

Therefore $\quad y_n + 4y_{n-1} - 21y_{n-2} = x_n + 3x_{n-1}$

and $\quad y_n = x_n + 3x_{n-1} - 4y_{n-1} + 21y_{n-2} .$

Taking $y_{-1} = y_{-2} = x_{-1} = 0$,

$$y_0 = x_0 = 1,$$

$$y_1 = x_1 + 3x_0 - 4y_0 = 1 + 3 - 4 = 0 ,$$

$$y_2 = x_2 + 3x_1 - 4y_1 + 21y_0$$

$$= 1 + 3 - 0 + 21 = 25 , \quad \text{etc.}$$

Example 6.5

Determine the impulse response sequence of the digital system with transfer function $\dfrac{5z^2}{4z^2 + 1}$.

The input to the system is $\{\delta_n\} = \{1, 0, 0, 0, \ldots\}$ and $Z\{\delta_n\} = 1$.

Therefore the impulse response sequence is given by

$$Z\{y_n\} = \left(\frac{5z^2}{4z^2 + 1}\right) . \ 1 = \frac{5z^2}{4z^2 + 1} .$$

Since Table 2A does not include a suitable entry for the inversion of this function of z we proceed by factorising the denominator into complex factors and applying the method of partial fractions, (after removing the factor z).

$$Z\{y_n\} = z \left(\frac{5z}{(2z - j)(2z + j)}\right)$$

$$= z\left(\frac{5/4}{2z - j} + \frac{5/4}{2z + j}\right) .$$

[Note: When performing partial fractions with complex conjugate linear factors, the numerator constants are *also* complex conjugates. Therefore once one of the constants has been determined the other may be written down directly.]

Continuing the example

$$Z\{y_n\} = \frac{5}{8}\left(\frac{z}{z - j/2}\right) + \frac{5}{8}\left(\frac{z}{z + j/2}\right) .$$

Therefore

$$\{y_n\} = \left\{\frac{5}{8}\left(\frac{j}{2}\right)^n + \frac{5}{8}\left(-\frac{j}{2}\right)^n\right\}$$

$$= \left\{\frac{5}{8}\left(\frac{1}{2}e^{j\,\pi/2}\right)^n + \frac{5}{8}\left(\frac{1}{2}e^{-j\,\pi/2}\right)^n\right\}$$

$$= \left\{\frac{5}{8}\left(\frac{1}{2}\right)^n e^{jn\,\pi/2} + \frac{5}{8}\left(\frac{1}{2}\right)^n e^{-jn\,\pi/2}\right\}$$

$$= \left\{\frac{5}{8}\left(\frac{1}{2}\right)^n 2\cos n\frac{\pi}{2}\right\}$$

$$= \left\{\frac{5}{4}\left(\frac{1}{2}\right)^n \cos n\frac{\pi}{2}\right\}$$

i.e. $$\{y_n\} = \left\{\frac{5}{4}\,,\,0\,,-\frac{5}{16}\,,\,0\,,\,\frac{5}{64}\,,\,0\,,\,\ldots\ldots\right\}.$$

The working in Example 6.5 may be shortened using the following general result which the reader should prove.

$$Z^{-1}\left\{z\left(\frac{\lambda+j\mu}{z-(\alpha+j\beta)}+\frac{\lambda-j\mu}{z-(\alpha-j\beta)}\right)\right\} = \{r^n(2\lambda\cos n\theta - 2\mu\sin n\theta)\}$$

where $\alpha + j\beta = re^{j\theta}$.

In Example 6.5, $\alpha = 0$, $\beta = \frac{1}{2}$, $r = \frac{1}{2}$, $\theta = \frac{\pi}{2}$, $\lambda = \frac{5}{8}$, $\mu = 0$.

Example 6.6

The function $f(t) = t$ is sampled at times $t = 0, 0.05, 0.1, 0.15, \ldots\ldots$, to produce the input to a system with transfer function $\frac{z-1}{z}$. Determine the output sequence from the system.

From Table 2B the input sequence has z-transform $\frac{0.05z}{(z-1)^2}$.

Therefore the output sequence $\{y_n\}$ is given by

$$Z\{y_n\} = \frac{0.05z}{(z-1)^2}\left(\frac{z-1}{z}\right) = \frac{0.05}{z-1}\,,$$

and from Table 2A

$$\{y_n\} = 0.05\,\{u(n-1)\} = 0.05\,\{0, 1, 1, 1, \ldots\}.$$

Problems

1. Solve the following difference equations to obtain the sequence $\{y_n\}$

(i)	$y_n = x_{n-2} + 5y_{n-1} - 6y_{n-2}$,	$\{x_n\} = \{1\}$
(ii)	$y_n = x_n - 2x_{n-1} + 3y_{n-1} - 2y_{n-2}$,	$\{x_n\} = \{1, 0, 0, 0, \ldots\}$
(iii)	$y_n = x_n - 4y_{n-1} - 4y_{n-2}$,	$\{x_n\} = \{2, 1, 0, 0, 0, \ldots\}$.

2. Write down the transfer functions relating the input and output of the systems described by the difference equations in question 1.

3. Solve the difference equation

$$f_{n+2} + 2f_{n+1} + f_n = 0 \,, f_0 = 1, \; f_1 = 0 \,, \; n = 0, 1, 2, 3, \ldots.$$

4. A discrete system has input sequence $\{x_n\}$, output sequence $\{c_n\}$ and transfer function $D(z) = \dfrac{4z^2}{4z^2 - 1}$. Write down the corresponding difference equation relating x_n and c_n. Obtain the sequence $\{c_n\}$ when $\{x_n\}$ is the unit impulse sequence $\{\delta_n\} = \{1, 0, 0, 0, \ldots\}$.

5. A discrete system has transfer function $\dfrac{5z^2}{4z^2 + 1}$. Determine the output sequence when the input sequence is the unit step sequence $\{u_n\} = \{1\}$.

6. A digital system is described by the transfer function $\dfrac{z(z + 1)}{(z + 2)^2}$. Obtain the output sequence for

    (i) a unit impulse sequence,
    (ii) a unit step sequence.

    Check your answers for $n = 0, 1, 2, 3$ by writing down the corresponding difference equation.

7. Obtain the output sequence of the system shown in Figure 6.2 when the sampler operates at $t = 0, 0.1, 0.2, 0.3, \ldots$.

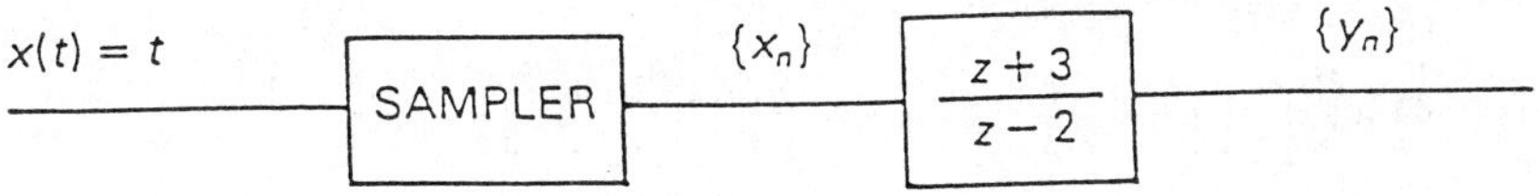

Fig. 6.2

8. A digital system, described by the transfer function $\dfrac{z(z + 1)}{(2z + 1)^2}$ has input sequence $\{x_n\}$ and output sequence $\{y_n\}$. Given that $x_0 = 4$, $x_n = 0$, $n = 1, 2, 3, \ldots$, express y_n in terms of n.

9. The output sequence $\{y_n\}$ of a certain digital processor satisfies the difference equation

$$y_n = x_n + x_{n-1} - 4y_{n-1} - 4y_{n-2}$$

where $y_n = 0$ for $n < 0$ and $\{x_n\}$, the input sequence, is defined by

$$x_n = \begin{cases} 2 \ , & n = 0 \\ 0 \ , & n \neq 0 \end{cases} .$$

Obtain an expression for y_n in terms of n.

10. Write down the z-transform of the sequence $\{x_n\}$ where

$$x_n = \begin{cases} 0 \ , & n < 0 \\ 2 \ , & n = 0 \\ 1 \ , & n = 1 \\ 0 \ , & n > 1 \end{cases} .$$

The output $\{y_n\}$ of a certain digital processor satisfies the difference equation

$$y_n = x_{n-1} - 3y_{n-1} - 2y_{n-2}$$

where $y_n = 0$ for $n < 0$ and $\{x_n\}$ is the sequence defined above. Obtain an expression for y_n in terms of n.

11. Show that

$$Z^{-1}\left\{\frac{z^2}{z^2 - 2z + 2}\right\} = \left\{(\sqrt{2})^n \left(\cos\frac{\pi n}{4} + \sin\frac{\pi n}{4}\right)\right\} .$$

Answers

1. (i) $\left\{\frac{1}{2} - 2^n + \frac{1}{2} . 3^n\right\}$ (ii) $\{1\}$ (iii) $\{2(-2)^n - 3n(-2)^{n-1}\}$.

2. (i) $\dfrac{1}{z^2 - 5z + 6}$ (ii) $\dfrac{z}{z - 1}$ (iii) $\dfrac{z^2}{z^2 + 4z + 4}$.

3. $\{f_n\} = \{(-1)^n (1 - n)\}$.

4. $4c_n = c_{n-2} + 4x_n$, $\{c_n\} = \left\{\left(\frac{1}{2}\right)^{n+1} - \left(-\frac{1}{2}\right)^{n+1}\right\}$.

5. $\left\{1 + \frac{1}{4}\left(\frac{1}{2}\right)^n \left(\cos n\frac{\pi}{2} + 2\sin n\frac{\pi}{2}\right)\right\}$.

6. (i) $\left\{(-2)^n\left(1 + \frac{n}{2}\right)\right\}$

   (ii) $\left\{\frac{2}{9} + \frac{(-2)^n}{9}(7 + 3n)\right\}$.

7. $\{y_n\} = \{-0.5 - 0.4n + 0.5(2)^n\}$.

8. $y_n = (1-n)\left(-\frac{1}{2}\right)^n$

9. $y_n = (2+n)(-2)^n$.

10. $Z\{x_n\} = 2 + z^{-1}$, $y_n = (3(-2)^{n-1} - (-1)^{n-1})\,u(n-1)$.

6.3 COMPARISON BETWEEN CONTINUOUS AND DISCRETE SYSTEMS

In Chapters 2 and 3 we saw how important results, such as stability and steady state solutions, could be obtained for continuous systems by consideration of the system transfer function without the necessity to obtain a complete solution of the corresponding differential equation. Similar techniques may be applied to digital systems once the system transfer function H(z) has been established, and information about system stability and steady solution can be obtained without solving the system difference equation.

The techniques depend on the relationship defined earlier between parameters s and z

$$z = e^{sT} \tag{6.2}$$

where the input sequence $\{x_n\}$ is considered to have been obtained by sampling some function $x(t)$ at times $t = 0$, T, 2T, ...

i.e. $$\{x_n\} = \{x(nT)\}.$$

Consider, for example, the digital system with transfer function

$$H(z) = \frac{3z^2 + 2z}{2z^2 + 5z - 3} .$$

The output of the system will be obtained from

$$Z\{y_n\} = \left(\frac{3z^2 + 2z}{(2z-1)(z+3)}\right) Z\{x_n\}$$

where $\{x_n\}$ is the input sequence.

Clearly $Z\{y_n\}$ will contain terms depending on the input (the "steady" solution) and terms $\frac{z}{z-0.5}$, $\frac{z}{z+3}$ arising from the system. These latter terms cause the output sequence to include $(0.5)^n$ and $(-3)^n$. Now $(0.5)^n \to 0$ as $n \to \infty$ but $(-3)^n$ is unbounded as $n \to \infty$. This particular system is said to be *unstable* because the part of the output sequence arising from the system itself (the "transient" solution) does not tend to a finite limit as $n \to \infty$.

In general if H(z) contains a *pole* at $z = a$ then $Z\{y_n\}$ will contain a term $\frac{z}{z-a}$ and y_n will include a^n. For stability *all* poles of H(z) must be such that $|z| < 1$, (i.e. $a^n \to 0$ as $n \to \infty$).

The *stability region* of the complex z-plane is shown in Figure 6.3 together with the stability region of the complex s-plane applicable to continuous systems (see Section 2.6). The relationship between the two regions arises directly from equation (6.2).

Writing $s = \sigma + jw$ the continuous system stability region is $\sigma < 0$.

Now $z = x + jy = e^{sT} = e^{\sigma T}(\cos wT + j \sin wT)$, and $|z|^2 = x^2 + y^2 = e^{2\sigma T}$.

Therefore $\sigma < 0$ corresponds to $|z| < 1$, the digital system stability region.

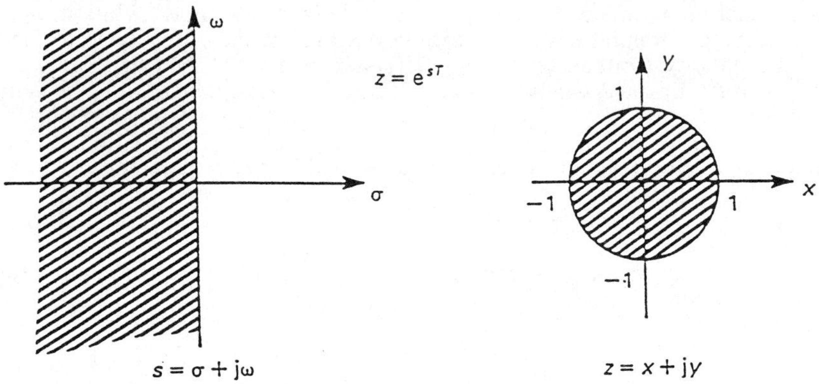

Fig. 6.3 Stability regions (shaded) for s-plane and z-plane.

Using the notation of Section 2.6 the output sequence of a digital system is given by

$$\{y_n\} = \{y_n\}_{TR} + \{y_n\}_{SS}$$

where $\{y_n\}_{TR}$ is the **transient response** of the system and contains terms independent of the input sequence $\{x_n\}$. For a *stable* system $\{y_n\}_{TR} \to 0$ as $n \to \infty$. $\{y_n\}_{SS}$ is the **steady state** solution which depends on the input sequence and for a stable system $\{y_n\} \to \{y_n\}_{SS}$ as $n \to \infty$.

For example, consider the system $H(z) = \dfrac{z+1}{\left(z - \frac{1}{2}\right)\left(z + \frac{1}{3}\right)}$ with input $\{x_n\} = \{u_n\}$, the unit step sequence.

The output sequence is given by

$$Z\{y_n\} = \frac{z(z+1)}{(z-1)\left(z - \frac{1}{2}\right)\left(z + \frac{1}{3}\right)}$$

$$= \frac{3z}{(z-1)} - \frac{18z}{5\left(z-\frac{1}{2}\right)} + \frac{3z}{5\left(z+\frac{1}{3}\right)} .$$

Therefore $\quad \{y_n\} = \left\{3 - \frac{18}{5}\left(\frac{1}{2}\right)^n + \frac{3}{5}\left(-\frac{1}{3}\right)^n\right\}$.

Here $\quad \{y_n\}_{TR} = \left\{-\frac{18}{5}\left(\frac{1}{2}\right)^n + \frac{3}{5}\left(-\frac{1}{3}\right)^n\right\}$

and $\quad \{y_n\}_{ss} = 3\{u_n\}$, (i.e. $y_n \to 3$ as $n \to \infty$).

In Section 2.6 we noted that y_{ss} for a continuous system with unit step function input was obtained immediately by putting $s = 0$ in the system transfer function. Similarly, for a digital system H(z), the steady value of the output for a unit step input sequence is obtained by putting $z = 1$ in H(z), since from equation (6.2) $z = e^{sT} = 1$ when $s = 0$.

Therefore, for $\{x_n\} = \{u_n\}$, $y_n \to$H(1) as $n \to \infty$, i.e. $\{y_n\}_{ss} = \text{H}(1)\{u_n\}$.

(This result can also be derived from the Final Value Theorem for z-transforms, property (ix) in Section 5.3.)

In addition, the first number in the output sequence for a unit step input sequence is given by $y_0 = \lim_{z \to \infty} \{\text{H}(z)\}$. This result follows immediately from the definition of the z-transform.

In the above example $\text{H}(z) = \dfrac{(z+1)}{\left(z-\frac{1}{2}\right)\left(z+\frac{1}{3}\right)}$ and, for $\{x_n\} = \{u_n\}$, $y_n \to \dfrac{2}{\frac{1}{2}\left(\frac{4}{3}\right)} = 3$ as $n \to \infty$. Also $y_0 = \text{H}(\infty) = 0$.

A similar short-cut applies when the input sequence is a sampled sine (or cosine) wave. In Chapter 3 for continuous systems we used the substitution $s = jw$. For digital systems we set $z = e^{jwT}$ in the system transfer function H(z). For the sampled input $\{x_n\} = \{a \sin wn\text{T}\}$ the steady state output sequence is given by

$$\{y_n\}_{ss} = \{a\text{R} \sin(wn\text{T} + \phi)\}$$

where R and ϕ are defined by

$$\text{H}(e^{jwT}) = \text{R}e^{j\phi} .$$

Summarising the results of this section applied to the system shown in Figure 6.4:

(i) The system is stable if all poles of H(z) lie **inside** the unit circle $|z| = 1$.

(ii) For $\{x_n\} = \text{K}\{u_n\}$, $\{y_n\}_{ss} = \text{KH}(1)\{u_n\}$, and $y_0 = \lim_{z \to \infty} \{\text{KH}(z)\}$.

(iii) For $x(t) = a \sin wt$, $\{y_n\}_{ss} = \{a\text{R} \sin(wn\text{T} + \phi)\}$, where $\text{H}(e^{jwT}) = \text{R}e^{j\phi}$.

All of these properties are related to the corresponding continuous system properties by application of equation (6.2). Table 3 on page 134 emphasises the similarity between continuous and discrete systems.

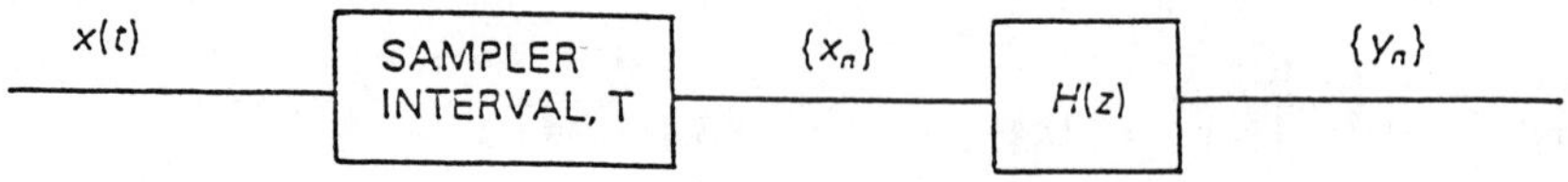

Fig. 6.4

Example 6.7

Suppose that T = 1 and $H(z) = \frac{z}{z - \frac{1}{2}}$ in Figure 6.4. Write down the steady state output sequence for the cases

(i) $\{x_n\} = 4\{u_n\}$,

(ii) $x(t) = \sin\frac{\pi}{3}t$.

(i) The input sequence is $4\{u_n\}$.

Therefore $\{y_n\}_{ss} = 4H(1)\{u_n\}$

$$= 8\{u_n\}.$$

(ii) The input sequence is $\left\{\sin\frac{\pi}{3}n\right\}$

i.e. $T = 1, w = \frac{\pi}{3}$ and we set $z = e^{j\frac{\pi}{3}}$.

Therefore $$H(e^{j\frac{\pi}{3}}) = \frac{e^{j\frac{\pi}{3}}}{e^{j\frac{\pi}{3}} - \frac{1}{2}}$$

$$= \frac{e^{j\frac{\pi}{3}}}{j\frac{\sqrt{3}}{2}}$$

$$= \frac{2e^{j\frac{\pi}{3}}}{\sqrt{3}\,e^{j\frac{\pi}{2}}} = \frac{2}{\sqrt{3}}e^{-j\frac{\pi}{6}}.$$

It follows that $R = \frac{2}{\sqrt{3}}$, $\phi = -\frac{\pi}{6}$ and $\{y_n\}_{ss} = \left\{\frac{2}{\sqrt{3}}\sin\left(\frac{\pi}{3}n - \frac{\pi}{6}\right)\right\}$.

Example 6.8

Given the difference equation

$$y_n + 2y_{n-1} + 2y_{n-2} = x_{n-1} + x_{n-2}\,, \quad x_n = y_n = 0 \text{ for } n < 0,$$

obtain the steady state solution for the case

$$\{x_n\} = \{2, 0, -2, 0, 2, 0, -2, 0, \ldots\}.$$

The difference equation can be written in the form

$$(1 + 2z^{-1} + 2z^{-2})Z\{y_n\} = (z^{-1} + z^{-2})\,Z\{x_n\}\,,$$

i.e.
$$H(z) = \frac{Z\{y_n\}}{Z\{x_n\}} = \frac{z+1}{z^2 + 2z + 2}$$

is the corresponding system transfer function.

The input sequence can be considered as the function

$$x(t) = 2\cos t \text{ sampled at}$$

$$t = 0\,,\ \frac{\pi}{2}\,,\ \pi\,,\ \frac{3\pi}{2}\,,\ 2\pi\,,\ \ldots\ldots,\ \text{i.e. } T = \frac{\pi}{2}\,,\ w = 1.$$

Therefore $\quad H(e^{jwT}) = H(j) = \frac{1+j}{1+2j} = Re^{j\phi}$ where $R = \sqrt{\frac{2}{5}}$, $\phi = \frac{\pi}{4} - \tan^{-1} 2.$

It follows that $\quad \{y_n\}_{ss} = \left\{2\sqrt{\frac{2}{5}}\cos\left(n\frac{\pi}{2} + \frac{\pi}{4} - \tan^{-1} 2\right)\right\}$

TABLE 3

Comparison between Continuous and Discrete Systems

<table>
<tr><th></th><th>CONTINUOUS</th><th>DISCRETE</th></tr>
<tr><td>NOTATION</td><td>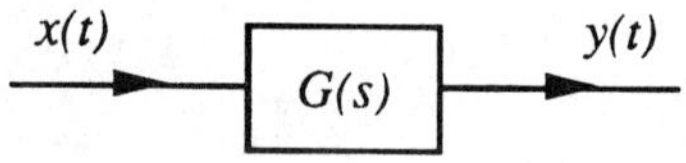

$L\{y(t)\} = G(s) \,.\, L\{x(t)\}$</td><td>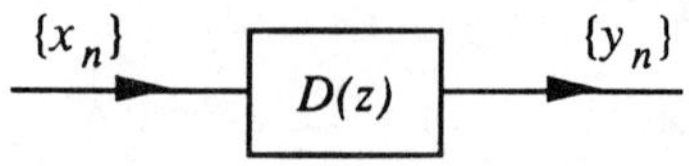

$Z\{y_n\} = D(z) \,.\, Z\{x_n\}$</td></tr>
<tr><td>DELAYED FUNCTION</td><td>$x(t\text{-}T)\, u(t\text{-}T) = \begin{cases} 0 \;, t < T \\ x(t\text{-}T), \; t \geq T \end{cases}$
$L\{x(t\text{-}T)\, u(t\text{-}T)\} = e^{-sT} L\{x(t)\}$</td><td>$\{x_{n\text{-}1}\} = (0, x_0, x_1, \ldots\}$
$Z\{x_{n\text{-}1}\} = z^{-1} Z\{x_n\}$</td></tr>
<tr><td>STABILITY</td><td>STABLE when <u>all</u> poles of $G(s)$ in left hand half of s-plane
$(\sigma < 0)$
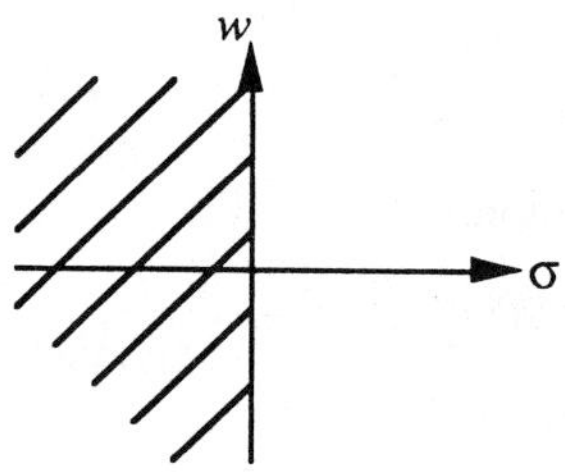

$s = \sigma + jw$</td><td>STABLE when <u>all</u> poles of $D(z)$ inside unit circle in z-plane
$(x^2+y^2 < 1)$
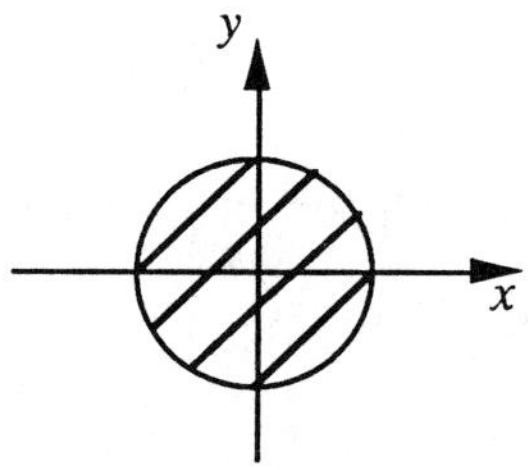

$z = x + iy$</td></tr>
<tr><td>INITIAL & FINAL VALUES (STEP INPUT)</td><td>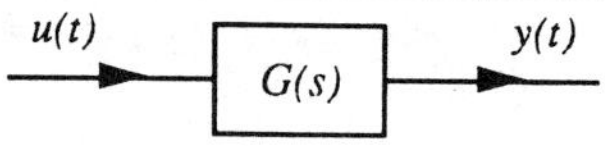

$y(0) = \lim_{s \to \infty} \{G(s)\}$
$\lim_{t \to \infty} \{y(t)\} = G(0)$</td><td>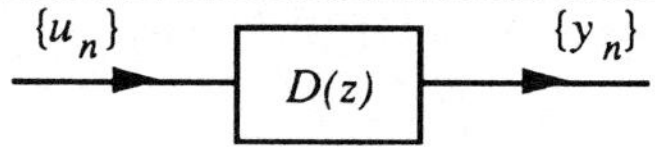

$y_0 = \lim_{z \to \infty} \{D(z)\}$
$\lim_{n \to \infty} \{y_n\} = D(1)$</td></tr>
<tr><td>FREQUENCY RESPONSE</td><td>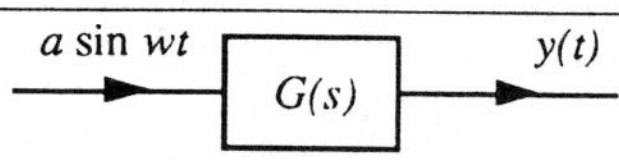

$y(t)_{ss} = aR \sin(wt + \phi)$
where $G(jw) = Re^{j\phi}$</td><td>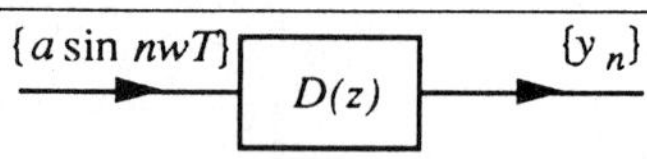

$\{y_n\}_{ss} = \{aR \sin(nwT + \phi)\}$
where $D(e^{jwT}) = Re^{j\phi}$</td></tr>
</table>

Problems

1. For the system shown in Figure 6.4 suppose that

$$T = \frac{\pi}{2} \quad \text{and} \quad H(z) = \frac{z}{4z^2 - 1} .$$

Confirm that the system is stable and obtain the steady state output sequence for the cases

(i) $x(t) = u(t)$, the unit step function,
(ii) $x(t) = 5 \sin t$.

2. Which of the following systems are stable?

(i) $$H(z) = \frac{z^2 + 3z + 2}{z(6z^2 + 5z + 1)} ,$$

(ii) $$H(z) = \frac{z + 1}{z^4 - 0.25} ,$$

(iii) $$H(z) = \frac{3z^2 + 2z}{4z^4 + 8z^3 + 9z^2 + 2z + 2} .$$

For the stable systems, calculate the steady state output sequence when the input sequence is the unit step sequence.

3. Consider the discrete system with transfer function $D(z) = \dfrac{z^2 + 2z}{3z^2 + 2z + 1}$, and output sequence $\{y_n\}$.

(i) Show that the system is stable.

(ii) Write down the steady-state output sequence when the input is a unit step sequence.

(iii) Write down the value of y_0 when the input is a unit step sequence.

4. The input sequence to the system D(z) is obtained by sampling $x(t) = a \sin wt$ at times $t = 0, T, 2T, 3T, \ldots$.

Determine the steady-state output sequence for the following situations, having confirmed the stability of D(z).

(i) $a = 2$, $w = 10\pi$, $T = 0.05$, $D(z) = \dfrac{2z - 1}{z + 0.5}$.

(ii) $a = 1$, $w = 1$, $T = 0.2$, $D(z) = \dfrac{z + 1}{11z - 9}$.

(iii) $a = 1$, $w = 100$, $T = \dfrac{\pi}{100}$, $D(z) = \dfrac{2z + 3}{z^2 + 0.25}$.

5. Obtain the steady state solution of the difference equation

$$y_n = \frac{1}{5}\left(2 \sin \frac{n\pi}{2} + 3y_{n-1}\right), \quad n = 0, 1, 2, 3, \ldots\ldots, \text{ where } y_{-1} = 0.$$

6. Write down the steady state output sequence for the digital system satisfying the difference equation.

$$2y_n = x_{n-3} + y_{n-1} - 0.25y_{n-3} + 0.125y_{n-4}\ , \quad n = 0\ , 1, 2, 3, \ldots$$

(where x_n and y_n are 0 for $n < 0$), given that $\{x_n\} = \{1, -1, 1, -1, 1, -1, \ldots\ldots\}$.

Answers

1. (i) $\frac{1}{3}\{u_n\}$ (ii) $\left\{\sin\,(n-1)\frac{\pi}{2}\right\}$.

2. (i) and (ii) are stable.

   (i) $\frac{1}{2}\{u_n\}$ (ii) $\frac{8}{3}\{u_n\}$.

3. (ii) $\left\{\frac{1}{2}\right\}$ (iii) $\frac{1}{3}$

4. (i) $\left\{4\sin\left(n\frac{\pi}{2} + 0.927\right)\right\}$

   (ii) $\{0.706\sin\,(n\,0.2 - 0.787)\}$

   (iii) $\{0\}$.

5. $\left\{\frac{2}{\sqrt{34}}\sin\left(n\frac{\pi}{2} - \tan^{-1} 0.6\right)\right\}$

6. $\left\{-\frac{8}{21}, \frac{8}{21}, -\frac{8}{21}, \frac{8}{21}, \ldots\ldots\right\}$

6.4 ALTERNATIVE METHODS FOR INVERSE Z-TRANSFORMS

In the previous sections, z-transforms were inverted using Table 2A after conversion to partial fraction form. Alternative methods shown in the following worked examples are

(i) power series inversion,
(ii) long division.

Example 6.9

Given that $F(z) = \frac{z+3}{z-2}$, find the sequence $\{f_n\}$ by

(i) using Table 2A,
(ii) power series inversion,
(iii) long division.

(i) $\quad F(z) = \dfrac{z}{z-2} + \dfrac{3}{z-2}$

Therefore $\quad \{f_n\} = \{2^n + 3(2)^{n-1}\, u(n-1)\}$,

i.e. $\quad f_0 = 1\ ;\ f_n = 5(2)^{n-1}\ ,\ n = 1, 2, 3, \ldots .$

(ii) $\quad F(z) = \dfrac{1+3z^{-1}}{1-2z^{-1}}$

$$= (1+3z^{-1})(1+2z^{-1}+2^2z^{-2}+2^3z^{-3}+\ldots\ldots)$$

$$= 1 + (3+2)z^{-1} + (3.2+2^2)z^{-2} + (3.2^2+2^3)z^{-3} + \ldots\ldots$$

$$= 1 + \sum_{n=1}^{\infty} (3(2)^{n-1} + 2^n)z^{-n} .$$

Therefore it follows from the definition of a z-transform that

$$f_0 = 1;\ f_n = 5(2)^{n-1}\ ,\ n = 1, 2, 3, \ldots\ldots .$$

(iii)

$$\begin{array}{r|l}
 & 1 + 5z^{-1} + 10z^{-2} + 20z^{-3} + \ldots\ldots\ldots \\
\hline
z-2 & z+3 \\
 & \underline{z-2} \\
 & \quad 5 \\
 & \quad 5 - 10z^{-1} \\
 & \quad \overline{\quad 10z^{-1}} \\
 & \quad\quad 10z^{-1} - 20z^{-2} \\
 & \quad\quad \overline{\quad\quad 20z^{-2}} \quad \text{etc.}
\end{array}$$

Therefore $\quad f_0 = 1, f_1 = 5, f_2 = 10, f_3 = 20\ , \ldots\ldots\ldots .$

This method is useful when only the first few numbers in the sequence are required, although in this example it is not difficult to obtain the general term.

Example 6.10

Given $\quad F(z) = \dfrac{z^3 - 9z^2 + 5z - 1}{4z^3 - 8z^2 + 5z - 1}$, find f_n for $n = 0, 1, 2$ using

(i) power series inversion,
(ii) long division.

(i) $\quad F(z) = \dfrac{z^3 - 9z^2 + 5z - 1}{(z-1)(2z-1)^2}$

$$= \frac{1 - 9z^{-1} + 5z^{-2} - z^{-3}}{4(1-z^{-1})(1-2^{-1}z^{-1})^2}$$

$$= \frac{1}{4}(1 - 9z^{-1} + 5z^{-2} - z^{-3})(1 + z^{-1} + z^{-2} + \ldots)\left(1 + \frac{2}{2}z^{-1} + \frac{3}{2^2}z^{-2} + \ldots\right)$$

$$= \frac{1}{4}\left(1 + \left(1 + \frac{2}{2} - 9\right)z^{-1} + \left(1 + \frac{3}{2^2} + 5 - 9 - 9 + \frac{2}{2}\right)z^{-2} + \ldots\ldots\right).$$

Therefore $\quad f_0 = \frac{1}{4}, f_1 = -\frac{7}{4}, f_2 = -\frac{41}{16}.$

(ii)

$$\begin{array}{r l} & \frac{1}{4} - \frac{7}{4}z^{-1} - \frac{41}{16}z^{-2} + \ldots\ldots \\ 4z^3 - 8z^2 + 5z - 1 \,| & z^3 - 9z^2 + 5z - 1 \\ & z^3 - 2z^2 + \frac{5}{4}z - \frac{1}{4} \\ \hline & -7z^2 + \frac{15}{4}z - \frac{3}{4} \\ & -7z^2 + 14z - \frac{35}{4} + \frac{7}{4}z^{-1} \\ \hline & -\frac{41}{4}z + 8 - \frac{7}{4}z^{-1} \end{array} \quad \text{etc.}$$

Therefore $\quad f_0 = \frac{1}{4}, f_1 = -\frac{7}{4}, f_2 = -\frac{41}{16}.$

For this example it is not easy to get a general form for f_n without using a partial fraction method.

Example 6.11

The input $\{x_n\}$ to a digital system with transfer function $\frac{z+1}{2z-1}$ is given by $x_n = n$. Obtain the output sequence $\{y_n\}$ using power series inversion.

$$Z\{x_n\} = \frac{z}{(z-1)^2}.$$

Therefore

$$Z\{y_n\} = \frac{z(z+1)}{(z-1)^2(2z-1)}$$

$$= \frac{z^{-1} + z^{-2}}{2(1 - z^{-1})^2(1 - 2^{-1}z^{-1})}$$

$$= \frac{1}{2}(z^{-1} + z^{-2})(1 + 2z^{-1} + 3z^{-2} + 4z^{-3} + \ldots)\left(1 + \frac{z^{-1}}{2} + \frac{z^{-2}}{2^2} + \frac{z^{-3}}{2^3} + \ldots\right)$$

$$= \frac{1}{2}(z^{-1} + z^{-2})\left(1 + \left(2 + \frac{1}{2}\right)z^{-1} + \left(3 + \frac{2}{2} + \frac{1}{2^2}\right)z^{-2} + \left(4 + \frac{3}{2} + \frac{2}{2^2} + \frac{1}{2^3}\right)z^{-3} + \ldots\right).$$

Therefore
$$y_0 = 0$$
$$y_1 = \frac{1}{2}$$
$$y_2 = \frac{1}{2}\left(1 + 2 + \frac{1}{2}\right)$$
$$y_3 = \frac{1}{2}\left(2 + \frac{1}{2} + 3 + \frac{2}{2} + \frac{1}{2^2}\right)$$
$$y_4 = \frac{1}{2}\left(3 + \frac{2}{2} + \frac{1}{2^2} + 4 + \frac{3}{2} + \frac{2}{2^2} + \frac{1}{2^3}\right)$$

i.e.
$$y_n = \frac{1}{2}\sum_{i=1}^{n} \frac{2i-1}{2^{n-i}} \quad \text{for } n \geq 2.$$

[This general form is not as convenient as the form obtained by the partial fraction approach, $y_n = 2n - 3 + 3\left(\frac{1}{2}\right)^n$. In general the power series approach (and the long division approach) is useful when the first few numbers are required and sometimes leads to a formula for the nth element of the sequence.]

Problems

1. Use power series inversion to show that

$$Z^{-1}\left\{\frac{1}{(z-a)^2}\right\} = \{(n-1)\, a^{n-2}\, u(n-1)\}\ .$$

Check the result of Example 6.10 by using partial fractions and transform tables to find a general formula for $f\,(n\mathrm{T})$.

2. Without using partial fractions find the inverse z-transform $\{f_n\}$ of the following F(z).

(i) $\dfrac{z^2+1}{z^2-1}$ (ii) $\dfrac{2z}{(z+1)(z+3)}$

(iii) $\dfrac{z^2+1}{(z-1)^3}$ (iv) $\dfrac{z}{z^2+1}$.

3. Given that the following F(z) were obtained by transforming $f(t)$, t = 0, T, 2T, 3T, ..., find the values $f(0)$, f (T), f (2T), f (3T) in each case.

(i) $\dfrac{z^3 + 3z^2 + 4z + 1}{(z-2)^3}$ (ii) $\dfrac{z^2+1}{(z+1)(z+2)^2}$

(iii) $e^{1/z}$ (iv) $\left(1 + \dfrac{1}{z}\right)^7$.

4. Use power series inversion to check the solutions of Problem 6, Section 6.2, for n = 0, 1, 2, 3.

5. Use long division to check the solution of Problem 7 in Section 6.2 for n = 0, 1, 2, 3.

Answers

2. (i) $f_0 = 1$; $f_n = 1 + (-1)^n$, $n = 1, 2, 3, \ldots$

   (ii) $f_n = (-1)^n - (-3)^n$, $n = 0, 1, 2, 3, \ldots$

   (iii) $f_0 = 0$; $f_n = n^2 - n + 1$, $n = 1, 2, 3, \ldots$

   (iv) $f_n = \sin \dfrac{n\pi}{2}$, $n = 0, 1, 2, 3, \ldots$

3. (i) 1, 9, 46, 177

   (ii) 0, 1, – 5, 18

   (iii) $1, 1, \frac{1}{2}, \frac{1}{6}$

   (iv) 1, 7, 21, 35.

7

Digital Filters

7.1 INTRODUCTION

The concept of filtering, i.e. shaping the frequency spectrum of an input signal, has its origin in the theory of linear continuous systems. A great deal of information is available for the design of a continuous filter (or **analog filter**) to meet a particular requirement. In this chapter we consider the problem of replacing a given analog filter by a suitably programmed digital computer operating on the sampled input. The motivation for this is that the cost of digital implementation is often considerably lower than that of its analog equivalent. A small special purpose computer can be time-shared among several components of a complex system and accuracy is dependent only on the computer word length. Also the filtering operation can be modified if necessary simply by changing the computer program, in contrast to the analog filter which may require to be physically rebuilt.

7.2 DIGITAL FILTER DESIGN

Figure 7.1(a) shows the analog filter G(s) together with its input and output, and Figure 7.1(b) shows its digital equivalent.

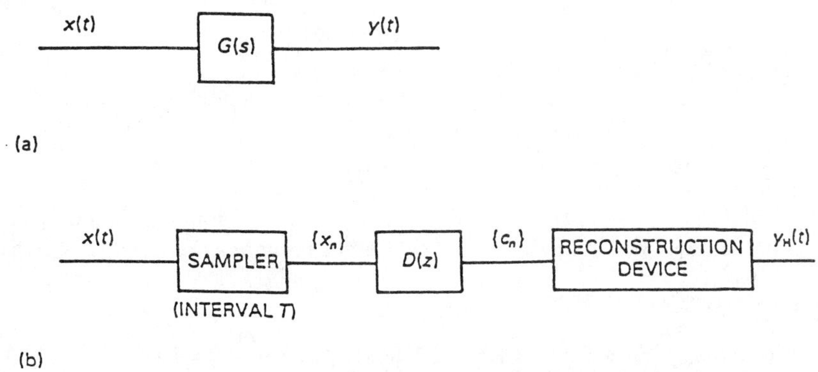

Fig. 7.1 (a) analog filter, (b) equivalent digital filter.

The sampler converts the continuous function $x(t)$ into the sequence $\{x_n\} = \{x(nT)\}$ and is sometimes referred to as an ADC (analog-to-digital converter). The reconstruction device is a form of DAC (digital-to-analog converter) which attempts to fit a continuous function $y_H(t)$ to the set of coordinates (nT, c_n), $n = 0, 1, 2, 3, \ldots$, where $\{c_n\}$ is the output sequence from the digital computer.

The simplest form of reconstruction device is the **zero-order hold** (ZOH) which outputs a "staircase" function as shown in Figure 7.2.

The transfer function D(z) describes the operation of the digital computer and is referred to as the **digital filter**. The object of this chapter is to design the digital filter D(z) so that the reconstructed output $y_H(t)$ is as close as possible to the analog filter output, $y(t)$. Once D(z) is known the digital computer can be programmed to execute the corresponding linear difference equation.

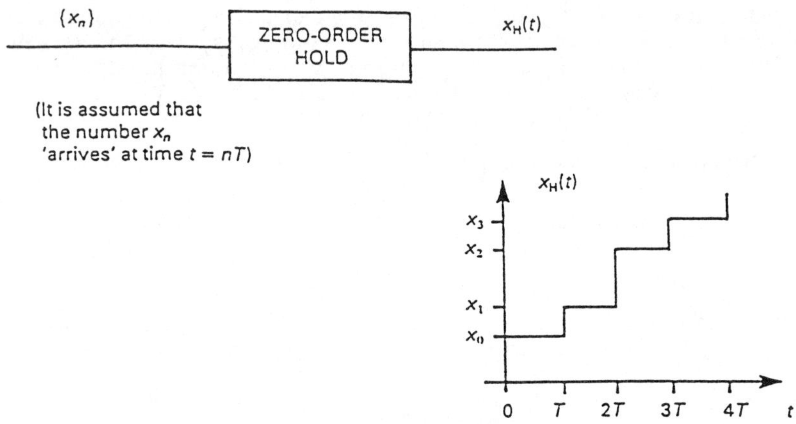

Fig. 7.2 Zero-order hold reconstruction device.

7.3 THE INPUT-INVARIANT TECHNIQUE

It is a simple matter to design D(z) such that the digital computer output sequence, $\{c_n\}$, satisfies $c_n = y(nT)$ for all *n provided that the input function, x(t),* is known.

The procedure is as follows:-

1. Calculate $y(t)$ from $y(t) = L^{-1}\{X(s)G(s)\}$, where $X(s) = L\{x(t)\}$.

2. Obtain the z-transforms $X(z) = Z\{x(t)\}$ and $Y(z) = Z\{y(t)\}$.

3. We require that $c_n = y(nT)$, i.e. $Z\{c_n\} = Z\{y(t)\}$.

   Therefore we take $D(z) = Y(z) / X(z)$ so that $Z\{c_n\} = X(z) . D(z) = Y(z)$.

If the reconstruction device is a zero-order hold then the outputs $y(t)$ and $y_H(t)$ compare as in Figure 7.3. Note that $y_H(t)$ lags behind $y(t)$ by an average amount T/2. This may not be significant if T is small, otherwise some advance needs to be built into the system.

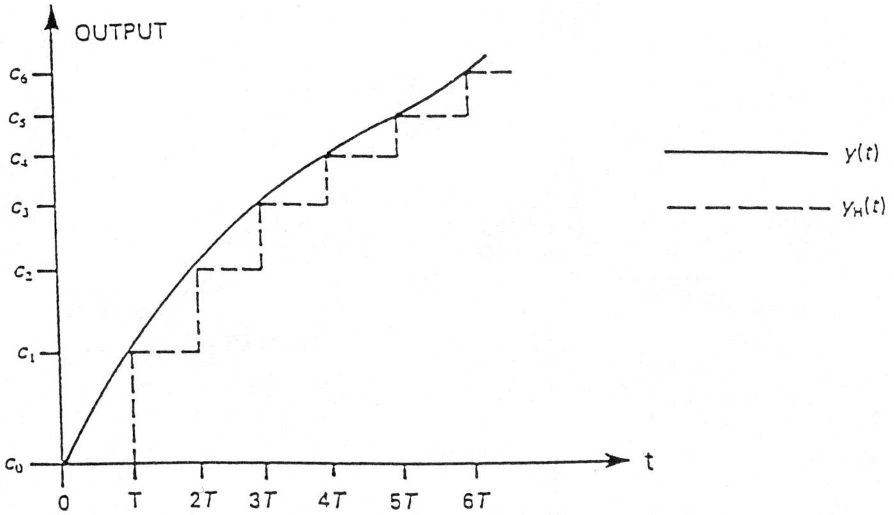

Fig. 7.3 Input-invariant technique with zero-order hold reconstruction.

The main disadvantage of the above technique is that $c_n = y(nT)$ exactly **only** for the particular input under consideration. The impulse-invariant design may give inaccurate output when the input is a step function, (see Example 7.1). Modifications to the technique are available to deal with inputs which are linear combinations of functions of the same type, see Stearns (1975). These are not included in this book, but the next section considers techniques which attempt to deal with general inputs.

Example 7.1

Design an impulse-invariant digital filter to be equivalent to the analog filter $\frac{1}{1+s}$. Compare the outputs of the analog and digital systems when the input is a unit step function.

The equivalent systems are shown in Figure 7.4.

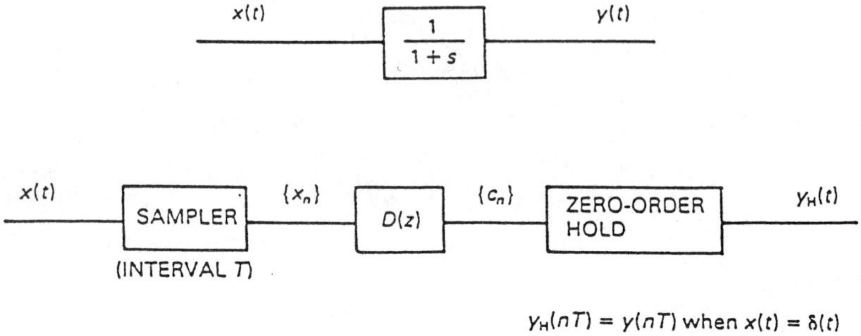

Fig. 7.4

For $x(t) = \delta(t)$, the unit impulse function, $L\{x(t)\} = 1$ and $L\{y(t)\} = \frac{1}{1+s}$.
Therefore $y(t) = e^{-t}\,u(t)$.

To analyse the digital system, we approximate to the unit impulse input by taking

$$x(t) = \begin{cases} \frac{1}{T} & \text{for } 0 \le t < T \\ 0 & \text{for } t \ge T \end{cases} .$$

Therefore $x_0 = \frac{1}{T}$, $x_n = 0$ for $n = 1, 2, 3, \ldots\ldots$ and $Z\{x_n\} = \frac{1}{T}$.

Also $\quad Z\{y(t)\} = \frac{z}{z - e^{-T}}$.

We require $\quad D(z) = \frac{Z\{y(t)\}}{Z\{x(t)\}} = \frac{Tz}{z - e^{-T}}$.

[We can check that $c_n = y(nT)$ for all n since

$$Z\{c_n\} = D(z) \, . \, Z\{x_n\} = \frac{z}{z - e^{-T}} .$$

Therefore $\quad c_n = e^{-nT} = y(nT)$.]

Now compare the outputs of the continuous system and the impulse-invariant digital system when the input is a unit step function.

i.e. $x(t) = u(t)$, $L\{x(t)\} = \frac{1}{s}$, $Z\{x(t)\} = \frac{z}{z-1}$.

For the analog system

$$L\{y(t)\} = \frac{1}{s(1+s)} = \frac{1}{s} - \frac{1}{1+s} \quad \text{and} \quad y(t) = (1 - e^{-t})\, u(t) .$$

For the impulse-invariant system

$$\begin{aligned} Z\{c_n\} &= \left(\frac{z}{z-1}\right)\left(\frac{Tz}{z-e^{-T}}\right) \\ &= \frac{T}{(1-e^{-T})}\left(\frac{z}{z-1} - \frac{e^{-T}z}{z-e^{-T}}\right) . \end{aligned}$$

Therefore

$$\begin{aligned} c_n &= \frac{T}{(1-e^{-T})}(1 - e^{-T}e^{-nT}) \\ &= \frac{T(1-e^{-(n+1)T})}{1-e^{-T}} \\ &= T\sum_{i=0}^{n} e^{-iT} . \end{aligned}$$

Comparing with $y(nT) = 1 - e^{-nT}$,

$y(0) = 0$	$c_0 = T$
$y(T) = 1 - e^{-T}$	$c_1 = T(1 + e^{-T})$
$Y(2T) = 1 - e^{-2T}$	$c_2 = T(1 + e^{-T} + e^{-2T})$
$\lim_{n\to\infty} y(nT) = 1$	$\lim_{n\to\infty} c_n = \frac{T}{1-e^{-T}}$.

For example if T = 0.2,

$y(0) = 0$	$c_0 = 0.2$
$y(0.2) = 0.1813$	$c_1 = 0.3637$
$y(0.4) = 0.3297$	$c_2 = 0.4978$
$\lim_{n\to\infty} y(0.2n) = 1$	$\lim_{n\to\infty} c_n = 1.1033$

This example illustrates that the impulse-invariant digital filter for $\frac{1}{1+s}$ does not perform particularly well for a step input, unless T is small. [Problem 1 suggests a method for modifying the impulse-invariant digital filter so that it gives more accurate results for a step-input.]

Example 7.2

Design a step-invariant digital filter to be equivalent to the analog filter $\frac{1}{1+s}$. Write down the difference equation to be executed by the digital computer.

$$x(t) = u(t)\ ,\ \mathrm{L}\{x(t)\} = \frac{1}{s}$$

Therefore $\qquad \mathrm{L}\{y(t)\} = \dfrac{1}{s(s+1)} = \dfrac{1}{s} - \dfrac{1}{1+s}$

and $\qquad y(t) = (1 - e^{-t})\, u(t).$

$$\mathrm{Z}\{x(t)\} = \frac{z}{z-1}$$

$$\mathrm{Z}\{y(t)\} = \frac{z}{z-1} - \frac{z}{z-e^{-T}}$$

Therefore take $\qquad \mathrm{D}(z) = \dfrac{\mathrm{Z}\{y(t)\}}{\mathrm{Z}\{x(t)\}} = 1 - \dfrac{z-1}{z-e^{-T}}$

$$= \frac{1-e^{-T}}{z-e^{-T}}\ .$$

To obtain the corresponding difference equation write

$$\frac{\mathrm{Z}\{c_n\}}{\mathrm{Z}\{x_n\}} = \mathrm{D}(z) = \frac{z^{-1}(1-e^{-T})}{1-z^{-1}e^{-T}}\ .$$

Therefore $\qquad c_n - e^{-T} c_{n-1} = (1 - e^{-T})\, x_{n-1}$

i.e. $\qquad c_n = (1 - e^{-T})\, x_{n-1} + e^{-T} c_{n-1}$

[Check $\qquad c_{-1} = 0\ ;\ x_{-1} = 0\ ,\ x_n = 1\ ,\ n \geq 0.$

$c_0 = 0\ ,\qquad$ compare $y(0) = 0$

$c_1 = 1 - e^{-T}\ ,\ $ compare $y(\mathrm{T}) = 1 - e^{-T},$

$c_2 = 1 - e^{-T} + e^{-T}(1 - e^{-T}) = 1 - e^{-2T}\ ,$

compare $y(2\mathrm{T}) = 1 - e^{-2T}\ ,$

$c_3 = 1 - e^{-T} + e^{-T}(1 - e^{-2T}) = 1 - e^{-3T}\ ,$

compare $y(3\mathrm{T}) = 1 - e^{-3T}$, etc.]

Problems

1. Example 7.1 designed the impulse-invariant digital filter $\mathrm{D}(z) = \dfrac{\mathrm{T}z}{z-e^{-T}}$ corresponding to the analog filter $\dfrac{1}{1+s}$. Modify this digital filter using a constant scale factor so that a *step* function input produces the same steady-state output for both analog and digital systems,

$$\text{(i.e.}\ \lim_{n\to\infty} c_n = \lim_{n\to\infty} y(n\mathrm{T})).$$

Compare the outputs when $\mathrm{T} = 0.2$.

2. Design a ramp-invariant digital filter (i.e. $x(t) = t$) to be equivalent to the analog filter $\frac{1}{1+s}$. Compare the analog and digital outputs when the input $x(t)$ is a unit step function and T = 0.2.

Answers

1. $$D(z) = \frac{(1-e^{-T})z}{z-e^{-T}}$$

$c_0 = 0.1813$, compare $y(0) = 0$

$c_1 = 0.3297$, compare $y(0.2) = 0.1813$

$c_2 = 0.4512$, compare $y(0.4) = 0.3297$

$c_n = 1 - e^{-0.2(n+1)}$ compare $y(0.2n) = 1 - e^{-0.2n}$

$\lim_{n \to \infty} c_n = 1$ compare $\lim_{n \to \infty} y(0.2n) = 1.$

2. $$D(z) = \frac{(T - 1 + e^{-T})z - Te^{-T} + 1 - e^{-T}}{T(z - e^{-T})}$$

For $x(t) = u(t)$, $c_n = 1 - \frac{1}{T}(e^{-nT} - e^{-(n+1)T})$

$$y(nT) = 1 - e^{-nT}$$

Comparing when T = 0.2

$c_0 = 0.0937$ $y(0) = 0$

$c_1 = 0.2579$ $y(0.2) = 0.1813$

$c_2 = 0.3925$ $y(0.4) = 0.3297$

$\lim_{n \to \infty} c_n = 1$, $\lim_{n \to \infty} y(0.2n) = 1.$

7.4 INPUT-INDEPENDENT TECHNIQUES

Many methods are available in the literature which attempt to overcome the difficulties of the previous section. Two such methods are considered in this section. The requirement is that the digital filter will operate satisfactorily for a range of input functions.

(i) The trapedoidal integration technique

Consider the analog system shown in Figure 7.5.

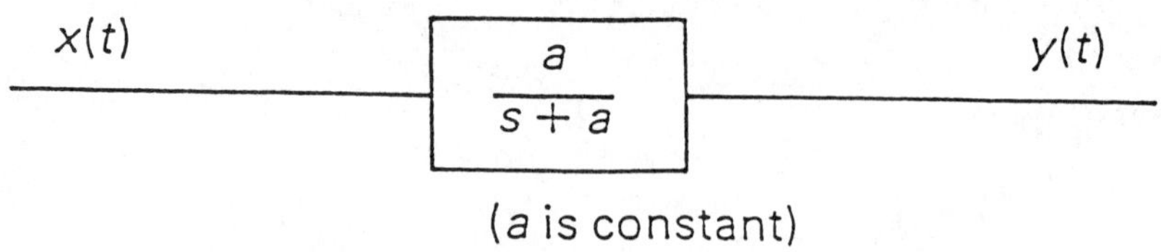

Fig. 7.5

The differential equation describing this system is

$$\frac{dy}{dt} + ay = ax ,$$

i.e.
$$\frac{dy}{dt} = a(x - y) .$$

We integrate this equation between $t = (n-1)\mathrm{T}$ and $t = n\mathrm{T}$ using trapezoidal integration.

$$\left[\int_{(n-1)T}^{nT} f(t)\, dt = \frac{\mathrm{T}}{2}(f_{n-1} + f_n) \ \text{where } f_n = f(n\mathrm{T}). \right]$$

$$\int_{(n-1)T}^{nT} dy = \int_{(n-1)T}^{nT} a(x-y)\, dt$$

i.e. $y_n - y_{n-1} = \dfrac{a\mathrm{T}}{2}(x_{n-1} - y_{n-1} + x_n - y_n)$.

[This equation can be solved explicitly for y_n and is the difference equation which is an approximation to the differential equation.]

Taking z-transforms

$$\left(1 - z^{-1} + \frac{a\mathrm{T}}{2}(1 + z^{-1})\right)\mathrm{Z}\{y_n\} = \frac{a\mathrm{T}}{2}(1 + z^{-1})\,\mathrm{Z}\{x_n\} .$$

Therefore
$$\frac{\mathrm{Z}\{y_n\}}{\mathrm{Z}\{x_n\}} = \frac{\frac{a\mathrm{T}}{2}(z+1)}{\frac{a\mathrm{T}}{2}(z+1) + (z-1)} .$$

We take this transfer function to be D(z), the digital filter, so that the digital output sequence $\{c_n\}$ is *approximately* $\{y_n\} = \{y(n\mathrm{T})\}$.

Therefore, for $\mathrm{G}(s) = \dfrac{a}{s + a}$, the trapezoidal approximation gives

$$D(z) = \frac{\frac{aT}{2}(z+1)}{\frac{aT}{2}(z+1) + (z-1)}$$

This can be written

$$D(z) = \frac{a}{a + \frac{2}{T}\left(\frac{z-1}{z+1}\right)} .$$

Note that D(z) is obtained from G(s) by the substitution $s = \frac{2}{T}\left(\frac{z-1}{z+1}\right)$.

Since any general (proper) transfer function, G(s), can be expressed by partial fractions as a combination of terms such as $\frac{a}{s+a}$, it follows that the substitution,

$$s = \frac{2}{T}\left(\frac{z-1}{z+1}\right)$$

can be applied to a *general* analog filter, G(s), to obtain the approximate corresponding digital filter, D(z).

The above substitution can be obtained using equation (6.2), $z = e^{sT}$.

$$z = e^{sT} = \frac{e^{sT/2}}{e^{-sT/2}}$$

$$= \frac{1 + sT/2}{1 - sT/2} \text{ , approximately, when T is small.}$$

Therefore $\quad sT/2 = \frac{z-1}{z-1} \quad$ and $\quad s = \frac{2}{T}\left(\frac{z-1}{z+1}\right)$.

Example 7.3

Use the trapezoidal approximation to design a digital filter corresponding to $G(s) = \frac{1}{s^2+s+1}$. Write down the difference equation relating output and input of the digital computer.

$$D(z) = \frac{1}{\left(\frac{2(z-1)}{T(z+1)}\right)^2 + \frac{2(z-1)}{T(z+1)} + 1}$$

$$= \frac{T^2(z+1)^2}{4(z-1)^2 + 2T(z^2-1) + T^2(z+1)^2}$$

$$= \frac{T^2(z^2+2z+1)}{(4+2T+T^2)z^2 + (2T^2-8)z + (4-2T+T^2)} .$$

i.e. $$\frac{Z\{c_n\}}{Z\{x_n\}} = \frac{T^2(1 + 2z^{-1} + z^{-2})}{(4 + 2T + T^2) + (2T^2 - 8)z^{-1} + (4 - 2T + T^2)z^{-2}} .$$

Therefore the required difference equation is

$$c_n = \frac{1}{(4 + 2T + T^2)}(T^2 (x_n + 2x_{n-1} + x_{n-2}) - (2T^2 - 8)c_{n-1} - (4 - 2T + T^2)c_{n-2}).$$

The trapezoidal approximation is very easy to use and has another advantage. The stable region in the complex s-plane is mapped into the stable region in the complex z-plane by the transformation $s = \frac{2}{T}\left(\frac{z-1}{z+1}\right)$. The stable region in the $s = \sigma + jw$ plane is $\sigma < 0$ and in Chapter 6 it was shown that the transformation $z = e^{sT}$ maps this region into $x^2 + y^2 < 1$ which is the stable region in the $z = x + jy$ plane, see Figure 7.6.

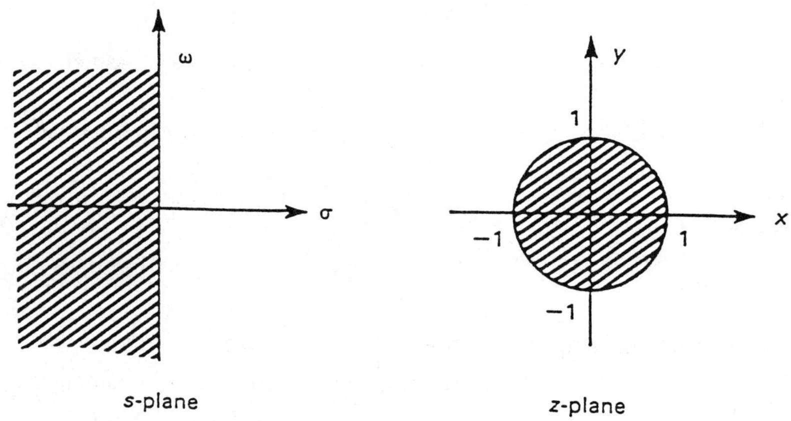

Fig. 7.6 Transformations $z = e^{sT}$ and $s = \frac{2}{T}\left(\frac{z-1}{z+1}\right)$.

It is easily proved that the trapezoidal transformation $s = \frac{2}{T}\left(\frac{z-1}{z+1}\right)$ carries out the same mapping. We have $\sigma + jw = \frac{2}{T}\left(\frac{(x-1) + jy}{(x+1) + jy}\right)$ and equating real parts gives

$$\sigma = \frac{2}{T}\left(\frac{x^2 + y^2 - 1}{(x+1)^2 + y^2}\right).$$

Therefore $\sigma < 0$ corresponds to $x^2 + y^2 < 1$.

It follows that, if the output of the analog system shown in Figure 7.1(a) is stable (i.e. G(s) has no poles in right hand half-plane) then the trapezoidal approximation for D(z) will produce *stable* digital computer output $\{c_n\}$, since D(z) will have no poles outside the unit circle.

(ii) The pole-zero matching technique

This method matches poles and zeros of G(s) and D(z) using the relation $z = e^{sT}$. For example, if G(s) has a pole at $s = -a$ then we would expect D(z) to have a pole at $z = e^{-aT}$. Before detailing this method we recall the trapezoidal approximation

$$D(z) = \frac{a}{a + \frac{2}{T}\left(\frac{z-1}{z+1}\right)} \quad \text{to} \quad G(s) = \frac{a}{a+s}\,.$$

After a little manipulation the trapezoidal D(z) can be written

$$D(z) = \left(\frac{a}{a + \frac{2}{T}}\right)\left(\frac{z+1}{z - \left(1 - \frac{aT}{2}\right) / \left(1 + \frac{aT}{2}\right)}\right).$$

If the sampling interval, T, is such that aT is small, we can write

$$\begin{aligned}\left(1 - \frac{aT}{2}\right)\left(1 + \frac{aT}{2}\right)^{-1} &\approx \left(1 - \frac{aT}{2}\right)\left(1 - \frac{aT}{2}\right)\\ &\approx 1 - aT\\ &\approx e^{-aT}\,.\end{aligned}$$

Therefore $D(z) \approx \left(\frac{a}{a + \frac{2}{T}}\right)\left(\frac{z+1}{z - e^{-aT}}\right)$ for the trapezoidal method. Thus the pole at $s = -a$ in G(s) shows up (as we would expect) as a pole at $z = e^{-aT}$ in D(z).

We still have to explain the presence of the $z + 1$ factor in the numerator. This corresponds to the fact the G(s) has a single zero when s is infinite, and the explanation is related to the choice of sampling interval, T. We consider the effect of the filter on high frequency inputs. The analog filter $\frac{a}{s+a}$ reduces towards zero input components of high frequency since $G(jw) = \frac{1}{a + jw} \to 0$ as $w \to \infty$.

It is reasonable to make the digital filter reduce to zero its maximum frequency input component. Assuming that the input $x(t)$ is band-limited with maximum frequency w_M rads/sec then the sampling interval T must be less than $\frac{\pi}{w_M}$ (Sampling Theorem, see Stearns (1975)). Put another way, for a given sampling interval T, the maximum frequency component of $x(t)$ must have frequency less than $\frac{\pi}{T}$ (Nyquist frequency). Therefore we make the digital filter reduce to zero, input with frequency $w = \frac{\pi}{T}$ i.e., D(z) must be zero when $s = j\frac{\pi}{T}$ and $z = e^{j\frac{\pi}{T}.T} = -1$. The pole-zero matching method is carried out as follows.

1. Any zero, $s = -b$, of G(s) implies that D(z) has a zero at $z = e^{-bT}$.

2. Any pole, $s = -a$, of G(s) implies that D(z) has a pole at $z = e^{-aT}$.

3. Any zero for s infinite implies that D(z) has a zero at $z = -1$.

4. Choose the gain of D(z) to match the gain of G(s) at some selected frequency, (*usually* $w = 0$, i.e. we match *steady-state* gains by choosing G(0) = D(1)).

For example if $G(s) = \dfrac{2}{(s+1)(s+2)}$ then

$$D(z) = \frac{(1-e^{-T})(1-e^{-2T})(z+1)^2}{4(z-e^{-T})(z-e^{-2T})}$$

(*double* zero for s infinite).

Example 7.4

Design a digital filter equivalent to $G(s) = \dfrac{1}{1+s}$ using

(i) trapezoidal approximation
(ii) pole-zero matching.

Compare the results from (i) and (ii) with the analog filter output given that the input is a unit step function and that T = 0.2.

(i) Trapezoidal approximation $D(z) = \dfrac{1}{1 + \dfrac{2}{T}\left(\dfrac{z-1}{z+1}\right)}$,

i.e. $$D(z) = \frac{(z+1)}{(z+1) + \dfrac{2}{T}(z-1)}$$

When T = 0.2, $D(z) = \dfrac{z+1}{11z-9}$.

Therefore $\dfrac{Z\{c_n\}}{Z\{x_n\}} = \dfrac{z+1}{11z-9}$ where $\{c_n\}$ is digital computer output.

The corresponding difference equation is

$$c_n = \frac{1}{11}(x_n + x_{n-1} + 9c_{n-1}).$$

For the unit step input, $x_n = 1, n \geq 0$.

Therefore $c_0 = 0.0909$, $c_1 = 0.2562$, $c_2 = 0.3914$,
$c_3 = 0.5021$, $c_4 = 0.5926$, $c_\infty = 1$.

(ii) Pole-zero matching $\quad D(z) = \left(\frac{1-e^{-T}}{2}\right)\left(\frac{z+1}{z-e^{-T}}\right).$

When T = 0.2, $D(z) = 0.09063\left(\frac{z+1}{z-0.8187}\right)$.

The corresponding difference equation is

$$c_n = 0.09063\,(x_n + x_{n-1}) + 0.8187c_{n-1}\ .$$

For the unit step input, $x_n = 1, n \geq 0$.

Therefore $\quad c_0 = 0.0906\ ,\quad c_1 = 0.2555\ ,\quad c_2 = 0.3904\ ,$
$\quad c_3 = 0.5009\ ,\quad c_4 = 0.5914\ ,\quad c_\infty = 1.$

Comparing the values in (i) and (ii) with the analog filter output $y(t) = 1 - e^{-t}$,

$$y(0) = 0\ ,\quad y(0.2) = 0.1813\ ,\quad y(0.4) = 0.3297$$
$$y(0.6) = 0.4512\ ,\quad y(0.8) = 0.5507\ ,\quad y(\infty) = 1.$$

Example 7.5

Design a digital filter equivalent to $G(s) = \frac{1}{s^2+1}$ using

(i) the trapezoidal approximation,
(ii) pole-zero matching.

Sketch a graph which compares the outputs of (i) and (ii) with the analog filter output when the input is a unit step function and $T = \frac{\pi}{3}$.

The analog filter output is given by

$$y(t) = L^{-1}\left\{\frac{1}{s(s^2+1)}\right\} = L^{-1}\left\{\frac{1}{s} - \frac{s}{(s^2+1)}\right\}$$
$$= (1 - \cos t)\, u(t)\ .$$

i.e. $\quad y(nT) = 1 - \cos n\frac{\pi}{3}$ when $T = \frac{\pi}{3}$.

One cycle of this output will be plotted on the graph and compared with the outputs of the two digital filters.

(i) Trapezoidal approximation $D(z) = \dfrac{1}{\frac{4}{T^2}\left(\frac{z-1}{z+1}\right)^2 + 1}$

i.e. $$D(z) = \frac{\frac{T^2}{4}(z+1)^2}{(z-1)^2 + \frac{T^2}{4}(z+1)^2}$$

When $T = \frac{\pi}{3}$, $D(z) = \dfrac{0.27416(z^2 + 2z + 1)}{1.27416z^2 - 1.45169z + 1.27416}$.

The corresponding difference equation is

$$c_n = 0.21517\,(x_n + 2x_{n-1} + x_{n-2}) + 1.13933c_{n-1} - c_{n-2}\,.$$

Therefore $c_0 = 0.2152$, $c_1 = 0.8907$, $c_2 = 1.6603$,
$c_3 = 1.8616$, $c_4 = 1.3214$, $c_5 = 0.5048$,
$c_6 = 0.1142$, $c_7 = 0.4862$, $c_8 = 1.3006$.

(ii) Pole-zero matching $D(z) = \dfrac{(1 - e^{jT})(1 - e^{-jT})(z + 1)^2}{4(z - e^{jT})(z - e^{-jT})}$

i.e. $$D(z) = \frac{(2 - 2\cos T)(z + 1)^2}{4(z^2 - 2\cos T\,.\,z + 1)}$$

when $T = \frac{\pi}{3}$, $D(z) = \dfrac{(z + 1)^2}{4(z^2 - z + 1)}$.

The corresponding difference equation is

$$c_n = \frac{1}{4}(x_n + 2x_{n-1} + x_{n-2}) + c_{n-1} - c_{n-2}\,.$$

Therefore $c_0 = 0.25$, $c_1 = 1$, $c_2 = 1.75$,
$c_3 = 1.75$, $c_4 = 1$, $c_5 = 0.25$,
$c_6 = 0.25$, $c_7 = 1$, $c_8 = 1.75$,

and the sequence repeats itself.

Comparing with the analog output $y(t) = 1 - \cos t$,

$$y(0) = 0\ ,\quad y\left(\frac{\pi}{3}\right) = 0.5\ ,\quad y\left(\frac{2\pi}{3}\right) = 1.5\ ,$$

$$y(\pi) = 2\ ,\quad y\left(\frac{4\pi}{3}\right) = 1.5\ ,\quad y\left(\frac{5\pi}{3}\right) = 0.5\ ,$$

$$y(2\pi) = 0\ ,\quad y\left(\frac{7\pi}{3}\right) = 0.5\ ,\quad y\left(\frac{8\pi}{3}\right) = 1.5\ ,\ \text{etc.}$$

The outputs are compared in Figure 7.7.

The "staircase" functions shown in Figure 7.7 represent the outputs from a zero-order hold reconstruction device, ———————— for pole-zero, – – – – – – – – – for trapezoidal. Note that the pole-zero sequence *leads* the analog output but that this is compensated by the *lag* due to zero-order hold reconstruction. The trapezoidal results are less accurate, but $T = \frac{\pi}{3}$ is rather a large sampling interval for this problem.

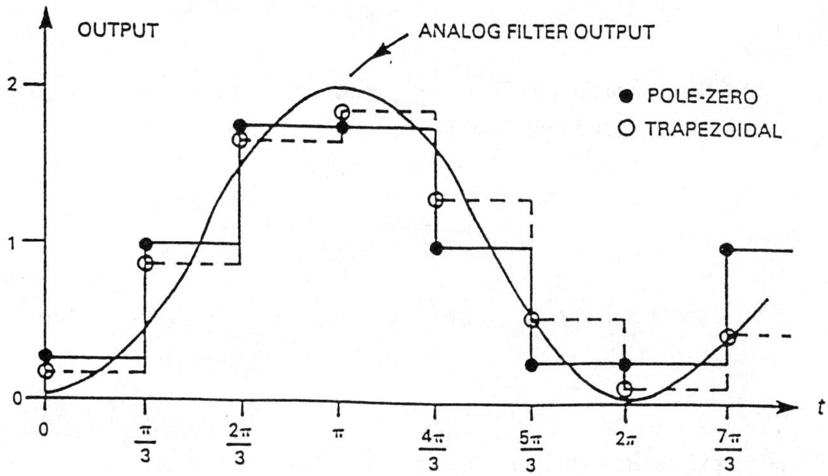

Fig. 7.7

Problems

1. Obtain the general formula for c_n in Example 7.4 for both trapezoidal integration and pole-zero matching.

2. Use the pole-zero matching method to write down the digital filters equivalent to

$$\text{(i)}\quad \frac{10}{s+10} \qquad \text{(ii)}\quad \frac{s+2}{(s+1)(s+3)} \qquad \text{(iii)}\quad \frac{10}{s^2+2s+5}\,.$$

3. Use the trapezoidal integration method to obtain digital filters equivalent to the analog filters in Problem 2.

4. (i) Use the pole-zero matching method to obtain a digital filter equivalent to $G(s) = \frac{1}{1+s}$ when the sampling interval is $T = \frac{\pi}{2}$. If the input function is sin t, find the general formula for the digital computer output sequence $\{c_n\}$. Deduce that as $n \to \infty$

$$|c_n| \leq 0.4586, \qquad \text{to 4 decimal place accuracy.}$$

   (ii) Use the trapezoidal approximation method to obtain a digital filter as in (i). Write down the corresponding difference equation relating digital computer output and input. For the input function, sin t, calculate c_n, $n = 0, 1, 2, 3, 4, 5$. Compare with the results in (i) and with the analog filter output.

5. Simpson's Rule integration procedure is as follows.

$$\int_{(n-2)T}^{nT} f(t)\,dt = \frac{T}{3}(f_{n-2} + 4f_{n-1} + f_n).$$

Apply this to the analog filter $\frac{a}{s+a}$ with input $x(t)$ and output $y(t)$ to show that the equivalent digital filter is

$$D(z) = \frac{a}{a + \frac{3}{T}\left(\frac{z^2 - 1}{z^2 + 4z + 1}\right)}.$$

Demonstrate that the transformation $s = \frac{3}{T}\left(\frac{z^2 - 1}{z^2 + 4z + 1}\right)$ does *not* map the stable region of the s-plane into the stable region of the z-plane.

6. The analog filter $G(s) = \frac{1 + 0.01s}{1 + 0.02s}$ is to be replaced by a digital filter $D(z)$ using the trapezoidal approximation. The input, $x(t)$, to $G(s)$ is sampled 200 times per second to form the sequence $\{x_n\}$ which is the input to $D(z)$. The output sequence from $D(z)$ is denoted by $\{c_n\}$.

(i) Obtain $D(z)$ and the difference equation relating $\{x_n\}$ and $\{c_n\}$.

(ii) For $x(t) = u(t)$ compare the initial and steady-state outputs of $G(s)$ and $D(z)$.

(iii) Compare the steady-state outputs of $G(s)$ and $D(z)$ when

$$x(t) = \sin 100\pi t.$$

7. Repeat Problem 6 for $G(s) = \frac{1}{1+s}$, $T = 0.2$, and with $x(t) = \sin t$ in part (iii).

8. A certain analog filter has input $x(t)$, output $y(t)$ and transfer function $G(s) = \frac{1}{(1 + 0.1s)^2}$. It is required to design an equivalent digital filter $D(z)$ whose input sequence $\{x_n\}$ is obtained by sampling $x(t)$, $\frac{1}{T}$ times per second. The output sequence $\{c_n\}$ of the digital filter is reconstructed to generate a continuous function.

(i) Write down the **steady-state** output of the analog filter when

$$x(t) = \sin 10t.$$

(ii) Use the trapezoidal method to design the digital filter $D(z)$, ensuring that $D(z)$ and $G(s)$ have the same steady-state output for $x(t) = u(t)$, the unit step function.

(iii) For the case $T = \frac{\pi}{20}$ determine the **steady-state** output sequence of the digital filter designed in part (ii) when $x(t) = \sin 10t$. Compare the gain and phase with the corresponding analog filter output.

(iv) Write down the initial outputs of G(s) and D(z) when $x(t) = u(t)$ and $T = \frac{\pi}{20}$.

9. The analog filter $G(s) = \frac{1 + 2\tau s}{1 + \tau s}$ is used to increase the phase in a control loop for an input frequency of 5 Hz. Show that this phase increase is maximum when $\tau = \frac{\sqrt{2}}{20\pi}$. Use the trapezoidal approximation to design a digital filter corresponding to G(s) with $\tau = \frac{\sqrt{2}}{20\pi}$, assuming that the input is sampled 20 times per second. Calculate the **steady-state** response of the digital system when the input to the sampler is sin $10\pi t$. Compare with the response of G(s) for this input.

10. An analog filter $G(s) = \frac{2}{2 + s}$ has input $x(t)$ and output $y(t)$. A digital filter D(z) equivalent to G(s) is required and the following design method is proposed.

Sample $x(t)$ 10 times per second and design D(z) using the trapezoidal method, ensuring that the steady-state output levels for G(s) and D(z) are equal for $x(t) = u(t)$.

Compare the steady-state outputs of D(z) and G(s) when $x(t) = \sin 5\pi t$.

Answers

1. (i) $c_n = 1 - \frac{10}{11}\left(\frac{9}{11}\right)^n$ (ii) $c_n = 1 - 0.9094\,(0.8187)^n$

2. (i) $\frac{(1 - e^{-10T})(z + 1)}{2(z - e^{-10T})}$ (ii) $\frac{(1 - e^{-T})(1 - e^{-3T})(z - e^{-2T})(z + 1)}{3(1 - e^{-2T})(z - e^{-T})(z - e^{-3T})}$

(iii) $\frac{(1 - 2e^{-T}\cos 2T + e^{-2T})(z + 1)^2}{2(z^2 - 2ze^{-T}\cos 2T + e^{-2T})}$

3. (i) $\frac{10T(z + 1)}{10T(z + 1) + 2(z - 1)}$ (ii) $\frac{2T^2(z + 1)^2 + 2T(z^2 - 1)}{3T^2(z + 1)^2 + 8T(z^2 - 1) + 4(z - 1)^2}$

(iii) $\frac{10T^2(z + 1)^2}{5T^2(z + 1)^2 + 4T(z^2 - 1) + 4(z - 1)^2}$.

4. (i) $\frac{0.39606(z + 1)}{z - 0.20788}$,

$$c_n = 0.3007 \sin\frac{n\pi}{2} - 0.4586 \cos\frac{n\pi}{2} + 0.4586(0.20788)^n .$$

(ii) $\frac{z + 1}{(z + 1) + 1.27324(z - 1)}$, $c_n = 0.4399\,(x_n + x_{n-1}) + 0.1202c_{n-1}$

$c_0 = 0$, $c_1 = 0.4399$, $c_2 = 0.4928$,
$c_3 = -0.3807$, $c_4 = -0.4857$, $c_5 = 0.3815$.

Compare with pole-zero results and analog output,

$$c_0 = 0 \quad , \quad c_1 = 0.3961 \quad , \quad c_2 = 0.4784 \, ,$$
$$c_3 = -0.2966 \quad , \quad c_4 = -0.4577 \quad , \quad c_5 = 0.3009.$$

$$y(0) = 0 \quad , \quad y\left(\frac{\pi}{2}\right) = 0.6039 \quad , \quad y(\pi) = 0.5216$$

$$y\left(\frac{3\pi}{2}\right) = -0.4955 \quad , \quad y(2\pi) = -0.4991 \, , \quad y\left(\frac{5\pi}{2}\right) = 0.5002.$$

6. (i) $D(z) = \dfrac{5z - 3}{9z - 7}$ (ii) $y(0) = 0.5 \, , \; c_0 = \dfrac{5}{9}$

$c_n = \dfrac{1}{9}(7c_{n-1} + 5x_n - 3x_{n-1})$ $y(\infty) = c_\infty = 1$

(iii) $y_{ss}(t) = 0.52 \sin(100\pi t - 0.15)$

$\{c_n\}_{ss} = \left\{0.51 \sin\left(n\frac{\pi}{2} - 0.12\right)\right\}.$

7. (i) $D(z) = \dfrac{z + 1}{11z - 9}$ (ii) $y(0) = 0 \, , \; c_0 = \dfrac{1}{11}$

$c_n = \dfrac{1}{11}(9c_{n-1} + x_n + x_{n-1})$ $y(\infty) = c_\infty = 1.$

(iii) $y_{ss}(t) = 0.707 \sin(t - 0.785)$

$\{c_n\}_{ss} = \{0.703 \sin(n0.2 - 0.79)\}$

8. (i) $y_{ss}(t) = \dfrac{1}{2}\sin\left(10t - \dfrac{\pi}{2}\right)$ (ii) $D(z) = \dfrac{T^2(z + 1)^2}{[(T + 0.2)z + (T - 0.2)]^2}$

(iii) $\{c_n\}_{ss} = \left\{0.381 \sin\left(\dfrac{n\pi}{2} - 1.81\right)\right\}$ (iv) $y(0) = 0 \, , \; c_0 = 0.1935.$

9. $D(z) = \dfrac{\pi(z + 1) + 4\sqrt{2}\,(z - 1)}{\pi(z + 1) + 2\sqrt{2}\,(z - 1)} = \dfrac{8.798z - 2.515}{5.970z + 0.313}$

$\{c_n\}_{ss} = \left\{1.53 \sin\left(\dfrac{n\pi}{2} + 0.33\right)\right\}.$

10. $D(z) = \dfrac{z + 1}{11z - 9}$

For D(z), $\{c_n\}_{ss} = \left\{0.0995 \sin\left(\dfrac{n\pi}{2} - 1.47\right)\right\}$

For G(s), $y_{ss}(n0.1) = 0.1263 \sin\left(\dfrac{n\pi}{2} - 1.444\right).$

7.5 FEEDBACK SYSTEMS WHICH INCLUDE A DIGITAL FILTER

A standard technique in continuous feedback control theory is to insert an analog filter in the forward path of the feedback system in an attempt to improve the stability of the system. The situation is shown in Figure 7.8(a) where the analog filter or "compensator" is denoted by $G_c(s)$. The analysis of the system is completed using the technique shown in Figure 2.5 to give

$$\frac{L\{y(t)\}}{L\{x(t)\}} = \frac{G_c(s)\,G(s)}{1 + G_c(s)\,G(s)}\,.$$

Once the input $x(t)$ is given, $y(t)$ may be determined using Laplace transforms.

Suppose now that the compensation is to be carried out using a digital computer, i.e. the analog filter $G_c(s)$ is replaced by a sampler/digital filter/ZOH combination as shown in Figure 7.8(b).

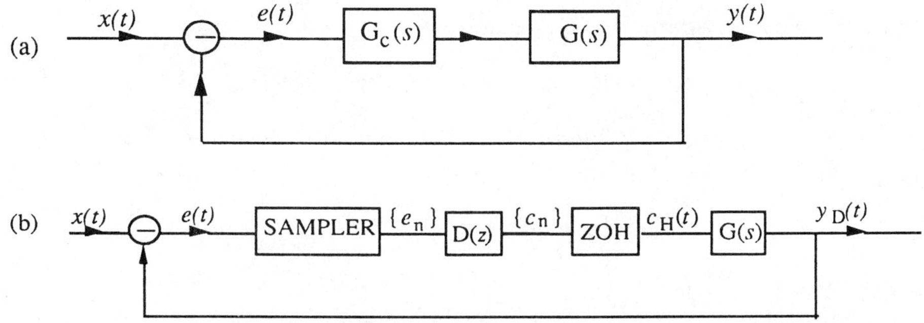

Fig. 7.8 Analog filter replaced by digital device.

The function $e(t)$ is the **error** i.e. the difference between input and output. The error is sampled to generate the sequence $\{e_n\} = \{e(nT)\}$ and the digital filter $D(z)$ implements a difference equation to generate the sequence $\{c_n\}$ which is reconstructed to form the function $c_H(t)$, the input to the component $G(s)$. It is assumed that the sampler and the zero-order hold both operate at the same sampling interval T.

In this section we attempt to determine the output $y_D(t)$ of the system with "digital compensation". The technique to be developed depends on the following important theorem.

Theorem 7.1

For the situation shown in Figure 7.9

$$Z\{y_D(t)\} = Z\{c_n\}\left(\frac{z-1}{z}\right)Z\left\{L^{-1}\left\{\frac{G(s)}{s}\right\}\right\}.$$

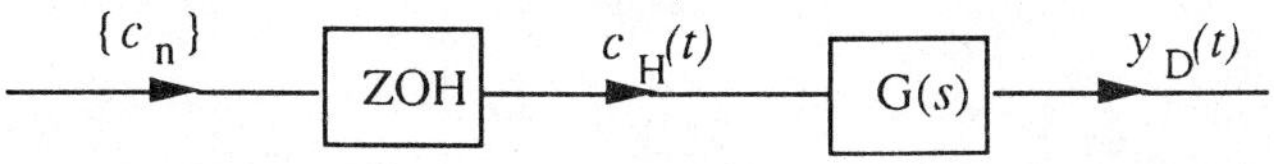

Fig. 7.9

Proof

With reference to Figure 7.2 we note that

$$c_H(t) = c_0 u(t) + (c_1 - c_0)\, u(t - T) + (c_2 - c_1)\, u(t - 2T) + \ldots\ldots$$

Therefore $\quad L\{c_H(t)\} = \dfrac{1}{s}\,(c_0 + (c_1 - c_0)e^{-sT} + (c_2 - c_1)e^{-s2T} + \ldots\ldots)$

and $\quad L\{y_D(t)\} = \dfrac{G(s)}{s}\,(c_0 + (c_1 - c_0)e^{-sT} + (c_2 - c_1)e^{-s2T} + \ldots\ldots).$

Denoting $\quad L^{-1}\left\{\dfrac{G(s)}{s}\right\}$ by $g_1(t)$ we obtain

$$y_D(t) = c_0 g_1(t) + (c_1 - c_0)\, g_1(t - T)u(t - T) + (c_2 - c_1)\, g_1(t - 2T)u(t - 2T) + \ldots\ldots$$

We now take z-transforms, recalling that

$$Z\{g_1(t - kT)\, u(t - kT)\} = z^{-k}\, Z\{g_1(t)\}\,, \quad \text{see Table 2B.}$$

$$\begin{aligned} Z\{y_D(t)\} &= Z\{g_1(t)\}\ (c_0 + (c_1 - c_0)z^{-1} + (c_2 - c_1)z^{-2} + \ldots\ldots) \\ &= Z\{g_1(t)\}\ (1 - z^{-1})(c_0 + c_1 z^{-1} + c_2 z^{-2} + \ldots\ldots) \\ &= Z\left\{L^{-1}\left\{\frac{G(s)}{s}\right\}\right\}\left(\frac{z-1}{z}\right) Z\{c_n\}\,, \quad \text{as required.} \end{aligned}$$

We note that the *sequence* $\{y_D(nT)\}$ is obtained from $Z^{-1}\{Z\{y_D(t)\}\}$.

Example 7.6

Suppose that in Figure 7.9 $G(s) = \dfrac{1}{s}$ and the zero-order hold operates with sampling interval $T = 0.2$. Further, the sequence $\{c_n\}$ is obtained from the output of a digital filter $D(z) = \dfrac{z+1}{11z-9}$ whose input is the unit step sequence $\{1, 1, 1, 1, \ldots\ldots\}$. Determine the sequence $\{y_D(n0.2)\}$.

$$\begin{aligned} Z\{c_n\} &= Z\{u_n\}\ D(z) \\ &= \left(\frac{z}{z-1}\right)\left(\frac{z+1}{11z-9}\right). \end{aligned}$$

Therefore $\quad Z\{y_D(t)\} = \left(\dfrac{z}{z-1}\right)\left(\dfrac{z+1}{11z-9}\right)\left(\dfrac{z-1}{z}\right) Z\left\{L^{-1}\left\{\dfrac{1}{s^2}\right\}\right\}$, from Theorem 7.1.

i.e. $\quad Z\{y_D(t)\} = \left(\dfrac{z+1}{11z-9}\right) Z\{t\}$

$$= \left(\frac{z+1}{11z-9}\right)\left(\frac{0.2z}{(z-1)^2}\right)$$

$$= \frac{z}{z - {}^{9}/_{11}} + \frac{0.2z}{(z-1)^2} - \frac{z}{z-1} .$$

$$\{y_D(n0.2)\} = Z^{-1}\{Z\{y_D(t)\}\}$$

$$= \left\{\left(\frac{9}{11}\right)^n + \frac{n}{5} - 1\right\} .$$

The result of Theorem 7.1 is now applied to the digitally compensated feedback system of Figure 7.8(b).

$$Z\{y_D(t)\} = Z\{c_n\}\left(\frac{z-1}{z}\right)Z\left\{L^{-1}\left\{\frac{G(s)}{s}\right\}\right\} \tag{7.1}$$

$$Z\{c_n\} = Z\{e(t)\}D(z) \tag{7.2}$$

$$Z\{e(t)\} = Z\{x(t)\} - Z\{y_D(t)\} \tag{7.3}$$

Elimination of $Z\{c_n\}$ from equations (7.1) and (7.2) gives

$$Z\{y_D(t)\} = Z\{e(t)\} . G(z) \tag{7.4}$$

where

$$G(z) = D(z)\left(\frac{z-1}{z}\right)Z\left\{L^{-1}\left\{\frac{G(s)}{s}\right\}\right\} \tag{7.5}$$

Further elimination of $Z\{e(t)\}$ from equations (7.3) and (7.4) leads to

$$Z\{y_D(t)\} = Z\{x(t)\}\left(\frac{G(z)}{1+G(z)}\right) \tag{7.6}$$

This final result is not surprising when we refer to Figure 2.5.

Given $x(t)$ we can use equation (7.6) to obtain $Z\{y_D(t)\}$ and the *sequence* $\{y_D(nT)\}$ is obtained from $Z^{-1}\{Z\{y_D(t)\}\}$. Note that the method of analysis developed in terms of z-transforms does not give the *function* $y_D(t)$. However, we can plot the coordinates $(nT, y_D(nT))$, $n = 0, 1, 2, 3, \ldots\ldots$, to obtain a graph of $y_D(t) \sim t$..

Example 7.7

Consider the simple feedback system shown in Figure 7.10.

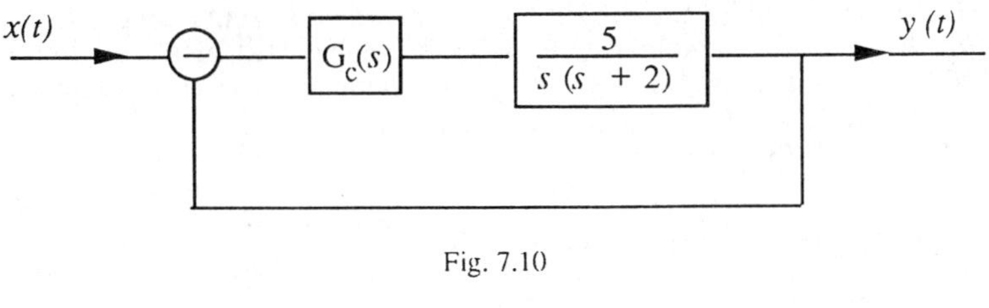

Fig. 7.10

It is given that $x(t) = u(t)$, the unit step function.

(i) For the case $G_c(s) = 1$, i.e. no compensation, determine the output $y(t)$.

(ii) Now take $G_c(s) = 0.4$, (compensation by a simple scaling factor) and determine the output $y(t)$. Comment on the improvement in system stability.

(iii) Replace the compensator $G_c(s) = 0.4$ by a sampler/digital filter/ZOH combination as in Figure 7.8(b). Given that the sampling interval is T = 0.5, determine the sequence $\{y_D(nT)\}$ for this digitally compensated system.

(i)
$$\frac{L\{y(t)\}}{L\{u(t)\}} = \frac{5/s\,(s+2)}{1+5/s(s+2)} = \frac{5}{s^2+2s+5}\,.$$

Therefore
$$L\{y(t)\} = \frac{5}{s(s^2+2s+5)}$$
$$= \frac{1}{s} - \frac{s+2}{s^2+2s+5}$$
$$= \frac{1}{s} - \frac{s+1}{(s+1)^2+2^2} - \frac{1}{(s+1)^2+2^2}$$

and
$$y(t) = \left(1 - e^{-t}\left(\cos 2t + \frac{1}{2}\sin 2t\right)\right)u(t)\,.$$

(ii)
$$\frac{L\{y(t)\}}{L\{u(t)\}} = \frac{2/s\,(s+2)}{1+2/s(s+2)} = \frac{2}{s^2+2s+2}\,.$$

Therefore
$$L\{y(t)\} = \frac{2}{s(s^2+2s+2)}$$
$$= \frac{1}{s} - \frac{s+2}{s^2+2s+2}$$
$$= \frac{1}{s} - \frac{s+1}{(s+1)^2+1^2} - \frac{1}{(s+1)^2+1^2}$$

and
$$y(t) = \left(1 - e^{-t}\,(\cos t + \sin t)\right)u(t)\,.$$

In both (i) and (ii) $y_{ss}(t) = 1$. In (i) the maximum value of $y(t)$ is approximately 1.21 whereas in (ii) this maximum value is reduced to about 1.04. Therefore (ii) is *less oscillatory* than (i), i.e. *more stable*.

(iii) Firstly we determine the digital filter $D(z)$ which is equivalent to $G_c(s) = 0.4$. For this simple example where $G_c(s)$ is independent of s we have, clearly, $D(z) = 0.4$. (Both trapezoidal and pole-zero methods will give this result.)

Therefore, from equation (7.5)

$$G(z) = 0.4\left(\frac{z-1}{z}\right)Z\left\{L^{-1}\left\{\frac{5}{s^2(s+2)}\right\}\right\}$$
$$= \left(\frac{z-1}{z}\right)Z\left\{L^{-1}\left\{\frac{1}{s^2} - \frac{0.5}{s} + \frac{0.5}{s+2}\right\}\right\}$$

$$= \left(\frac{z-1}{z}\right)Z\{t - 0.5 + 0.5e^{-2t}\}$$

$$= \left(\frac{z-1}{z}\right)\left(\frac{Tz}{(z-1)^2} - \frac{0.5z}{z-1} + \frac{0.5z}{z-e^{-2T}}\right), \text{ from Table 2B,}$$

$$= 0.5\left(\frac{1}{z-1} - 1 + \frac{z-1}{z-e^{-1}}\right), \text{ since T} = 0.5$$

$$= \frac{0.5(ze^{-1} - 2e^{-1} + 1)}{(z-1)(z-e^{-1})} .$$

Equation (7.6) is

$$Z\{y_D(t)\} = Z\{x(t)\}\left(\frac{G(z)}{1+G(z)}\right)$$

$$= \left(\frac{z}{z-1}\right)\left(\frac{0.5(ze^{-1} - 2e^{-1} + 1)}{(z-1)(z-e^{-1}) + 0.5\,(ze^{-1} - 2e^{-1} + 1)}\right),$$

since $x(t) = u(t)$,

$$= \left(\frac{z}{z-1}\right)\left(\frac{0.5(ze^{-1} + 1 - 2e^{-1})}{z^2 - z\,(1 + 0.5e^{-1}) + 0.5}\right) ,$$

$$\approx \frac{0.1839z^2 + 0.1321z}{z^3 - 2.1839z^2 + 1.6839z - 0.5} .$$

Now $y_D(n0.5) = Z^{-1}\{Z\{y_D(t)\}\}$ and we may use the long division inversion method described in Section 6.4. This gives

$$Z\{y_D(t)\} = 0.184z^{-1} + 0.534z^{-2} + 0.856z^{-3} + 1.062z^{-4} + 1.146z^{-5} + 1.141z^{-6} + 1.094z^{-7} + \ldots$$

Therefore $y_D(0) = 0$, $y_D(0.5) = 0.184$
$y_D(1) = 0.534$, $y_D(1.5) = 0.856$
$y_D(2) = 1.062$, $y_D(2.5) = 1.146$
$y_D(3) = 1.141$, $y_D(3.5) = 1.094$

etc.

The graph of $y_D(t) \sim t$ may now be drawn and we see that the maximum value of $y_D(t)$ is about 1.16. Thus the digitally compensated system is less stable than the analog system it represents. This is because the "delay" inherent in sampler/zero-order hold devices has a destabilising effect on the system.

Example 7.8

A feedback system is as shown in Figure 7.8(a) with compensator $G_c(s) = \frac{1+7s}{1+70s}$ and process transfer function $G(s) = \frac{40}{s(s+2)}$. Suppose that $G_c(s)$ is replaced by a sampler/digital filter/ZOH combination using the trapezoidal method to design D(z) based on a sampling interval T = 1. Determine the output $y_D(t)$ of the digitally compensated system at times $t = 0, 1, 2, 3, 4$ and 5.

[Clearly the sampling rate is too slow for a realistic system and the example is completed purely for demonstration.]

Using the trapezoidal approximation $s = \frac{2}{T}\left(\frac{z-1}{z+1}\right)$,

$$D(z) = \frac{1 + 14\left(\frac{z-1}{z+1}\right)}{1 + 140\left(\frac{z-1}{z+1}\right)} = \frac{15z - 13}{141z - 139}.$$

$$G(z) = D(z)\left(\frac{z-1}{z}\right) Z\left\{L^{-1}\left\{\frac{40}{s^2(s+2)}\right\}\right\}$$

$$= \left(\frac{15z-13}{141z-139}\right)\left(\frac{z-1}{z}\right)\left(\frac{20z}{(z-1)^2} - \frac{10z}{z-1} + \frac{10z}{z-e^{-2}}\right),$$

following similar working to Example (7.7).

Therefore $$G(z) = \left(\frac{15z-13}{141z-139}\right)\left(\frac{20}{z-1} - 10 + \frac{10(z-1)}{z-e^{-2}}\right)$$

$$= \left(\frac{15z-13}{141z-139}\right)\left(\frac{z(10+10e^{-2}) + 10 - 30e^{-2}}{(z-1)(z-e^{-2})}\right)$$

also $$\frac{Z\{y_D(t)\}}{Z\{x(t)\}} = \frac{G(z)}{1+G(z)} \quad \text{and} \quad L\{x(t)\} = \frac{z}{z-1}.$$

Therefore $$L\{y_D(t)\} \approx \frac{170.3z^3 - 58.49z^2 - 77.22z}{141z^4 - 269.78z^3 + 247.18z^2 - 214.43z + 96.03}$$

$$= 1.208z^{-1} + 1.896z^{-2} + 0.963z^{-3} + 0.355z^{-4} + 1.053z^{-5} + \ldots,$$

and $y(0) = 0$, $y(1) = 1.208$, $y(2) = 1.896$,
$y(3) = 0.963$, $y(4) = 0.355$, $y(5) = 1.053$, etc.

It appears that the modified system is converging very slowly, i.e. stability is unsatisfactory. A much smaller value for the sampling interval is required to obtain stability similar to that of the original system.

The reader is invited to check the long divisions in Examples 7.7 and 7.8.

Problems

1. The system shown in Figure 7.11 has a controller $G_c(s) = \dfrac{1}{1+s}$ which is to be replaced by a sampler/digital processor/zero-order hold combination operating with a sampling interval T = 1. Use the trapezoidal approximation to determine a digital filter equivalent to $\dfrac{1}{1+s}$ and calculate the output $y_D(t)$ of the modified system at times $t = 0, 1, 2, 3$ and 4.

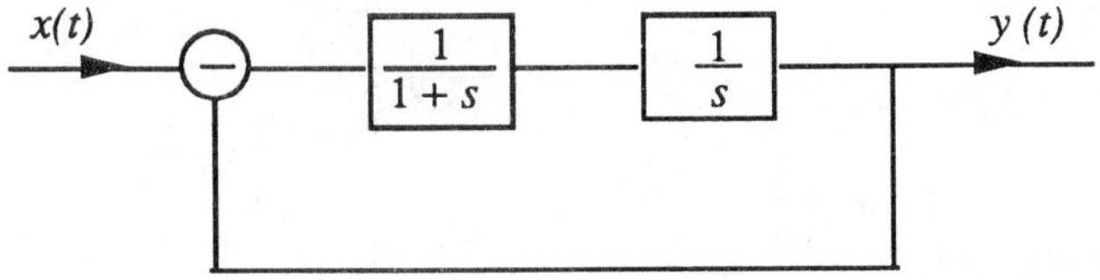

Fig. 7.11

2. A feedback system is as shown in Figure 7.8(a) with a controller $G_c(s) = 1 + \dfrac{2}{s}$ and a process transfer function $G(s) = \dfrac{1}{1+s}$. The controller is replaced by a digital device comprising a sampler/digital processor/zero-order hold combination operating at a sampling rate of two samples per second. Show that $D(z) = \dfrac{3z-1}{2(z-1)}$ is a suitable digital filter corresponding to $G_c(s)$ and calculate the output of the digitally controlled system at the times $t = 0, 0.5, 1.0, 1.5$ when the system input is $u(t)$ the unit step function.

Answers

1. $\dfrac{z+1}{3z-1}$; 0 , $\dfrac{1}{3}$, 1 , $\dfrac{13}{9}$, $\dfrac{13}{9}$.

2. 0 , 0.590 , 0.993 , 1.161.

REFERENCES

Apostol, T.M., (1974) *Mathematical analysis,* Addison Wesley

Braun, M., (1975) *Differential equations and their applications,* Springer-Verlag

Burghes, D.N., & Graham, A., (1980) *Introduction to control theory including optimal control,* Ellis Horwood

Gabel, R. A., & Roberts, R.A., (1973) *Signals and linear systems,* John Wiley

Holbrook, J.G., (1966) *Laplace transforms for electronic engineers,* Pergamon Press

Huntley, I.D., & Johnson, R.M., (1983) *Linear and nonlinear differential equations,* Ellis Horwood

Kreyszig, E., (1993) *Advanced engineering mathematics,* John Wiley

Oppenheim, A.V., & Wilsky, A.S., (1983) *Signals and systems,* Prentice-Hall

Stanley, W.D., (1975) *Digital signal processing,* Reston

Stearns, S.D., (1975) *Digital signal analysis,* Hayden

Index

Index

Albion Books on Mathematics and its Applications

MATHEMATICAL ANALYSIS AND PROOF

DAVID S.G. STIRLING, Senior Lecturer in Mathematics, University of Reading

ISBN 1-898563-36-5 250 pages 1997

This fundamental and straightforward text addresses a weakness observed among present-day students, namely a lack of familiarity with formal proof. This approach offers intellectual challenge and stimulus to readers by emphasising: (1) the need for familiarity with long mathematical arguments and manipulation, (2) the importance of the ability to construct proofs in analysis. The sample sections are impressive: the author teaches the reader (by example) to write in good mathematical style and elegant prose. The book is written in "easy" language, usefully explained with clarity, logic and brevity. Over 100 Worked Examples and more than 200 Problem Exercises (with solutions and hints). Index of Notation and Subject Index

Contents: 1 Setting the Scene 2 Logic and Deduction 3 Mathematical Induction 4 Sets and Numbers 5 Order and Inequalities 6 Decimals 7 Limits 8 Infinite Series 9 Structure of Real Number System 10 Continuity 11 Differentiation 12 Functions Defined by Power Series 13 Integration 14 Functions of Several Variables; Appendix: Decimal Expansion of Integers References

TEACHING AND LEARNING MATHEMATICAL MODELLING

Editors: S.K. HOUSTON, University of Ulster, Northern Ireland; W. BLUM, University of Kassel, Germany; IAN HUNTLEY, University of Bristol, England; N.T. NEILL, University of Ulster, Northern Ireland

ISBN- 1-898563-29-2 416 pages 1997

This contribution is sponsored by the organising committee of International Conferences on the Teaching of Mathematical Modelling and Applications (ICTMA). Mathematicians from Austria, Australia, Germany, Holland, Italy, Japan, Russia, Spain, UK, and USA contribute to teaching mathematical modelling in universities, technical colleges, teacher training colleges and high schools.

The interdisciplinary nature of the topics reflects their use in such varied areas of application as mechanics and engineering, medicine, patient flow through hospitals, computing science, traffic control, business studies, and mathematics (fractals and analysis), all pointing to a wide choice of future careers.

SECTIONAL CONTENTS: Section A Reflections and Investigations; *Section B* Assessment at Undergraduate Level; *Section C* Secondary courses and Case Studies; *Section D* Tertiary Case Studies; *Section E* Tertiary Courses.

SIGNAL PROCESSING IN ELECTRONIC COMMUNICATIONS

MICHAEL J. CHAPMAN, DAVID P. GOODALL, and NIGEL C. STEELE, School of Mathematics and Information Sciences, Coventry University

ISBN 1-898563-23-3 *ca.* 300 pages 1997

This text for advanced undergraduate and postgraduate courses in electronic engineering, applied mathematics, and computer science, deals with signal processing as an important aspect of electronic communications in its role of transmitting information, and the language of its expression. It develops the required mathematics in an interesting and informative way.

Contents: 1 Signal and Linear System Fundamentals 2 System Responses 3 Fourier Methods 4 Analogue Filters 5 Discrete-time Signals and Systems 6 Discrete-time System Responses 7 Discrete-time Fourier Analysis 8 The Design of Digital Filters 9 Aspects of Speech Processing Appendixes: (a) The Complex Exponential (b) Linear Predictive Coding Algorithms (c) Answers to the exercises.